Contents

Logarithms

	0	1	2	3	4	5	6	7	8	9	Mean Differences								
											1	2	3	4	5	6	7	8	9
10	0000	0043	0086	0128	0170	0212	0253	0294	0334	0374	4	8	12	17	21	25	29	33	37
11	0414	0453	0492	0531	0569	0607	0645	0682	0719	0755	4	8	11	15	19	23	26	30	34
12	0792	0828	0864	0899	0934	0969	1004	1038	1072	1106	3	7	10	14	17	21	24	28	31
13	1139	1173	1206	1239	1271	1303	1335	1367	1399	1430	3	6	10	13	16	19	23	26	29
14	1461	1492	1523	1553	1584	1614	1644	1673	1703	1732	3	6	9	12	15	18	21	24	27
15	1761	1790	1818	1847	1875	1903	1931	1959	1987	2014	3	6	8	11	14	17	20	22	25
16	2041	2068	2095	2122	2148	2175	2201	2227	2253	2279	3	5	8	11	13	16	18	21	24
17	2304	2330	2355	2380	2405	2430	2455	2480	2504	2529	2	5	7	10	12	15	17	20	22
18	2553	2577	2601	2625	2648	2672	2695	2718	2742	2765	2	5	7	9	12	14	16	19	21
19	2788	2810	2833	2856	2878	2900	2923	2945	2967	2989	2	4	7	9	11	13	16	18	20
20	3010	3032	3054	3075	3096	3118	3139	3160	3181	3201	2	4	6	8	11	13	15	17	19
21	3222	3243	3263	3284	3304	3324	3345	3365	3385	3404	2	4	6	8	10	12	14	16	18
22	3424	3444	3464	3483	3502	3522	3541	3560	3579	3598	2	4	6	8	10	12	14	15	17
23	3617	3636	3655	3674	3692	3711	3729	3747	3766	3784	2	4	6	7	9	11	13	15	17
24	3802	3820	3838	3856	3874	3892	3909	3927	3945	3962	2	4	5	7	9	11	12	14	16
25	3979	3997	4014	4031	4048	4065	4082	4099	4116	4133	2	3	5	7	9	10	12	14	15
26	4150	4166	4183	4200	4216	4232	4249	4265	4281	4298	2	3	5	7	8	10	11	13	15
27	4314	4330	4346	4362	4378	4393	4409	4425	4440	4456	2	3	5	6	8	9	11	13	14
28	4472	4487	4502	4518	4533	4548	4564	4579	4594	4609	2	3	5	6	8	9	11	12	14
29	4624	4639	4654	4669	4683	4698	4713	4728	4742	4757	1	3	4	6	7	9	10	12	13
30	4771	4786	4800	4814	4829	4843	4857	4871	4886	4900	1	3	4	6	7	9	10	11	13
31	4914	4928	4942	4955	4969	4983	4997	5011	5024	5038	1	3	4	6	7	8	10	11	12
32	5051	5065	5079	5092	5105	5119	5132	5145	5159	5172	1	3	4	5	7	8	9	11	12
33	5185	5198	5211	5224	5237	5250	5263	5276	5289	5302	1	3	4	5	6	8	9	10	12
34	5315	5328	5340	5353	5366	5378	5391	5403	5416	5428	1	3	4	5	6	8	9	10	11
35	5441	5453	5465	5478	5490	5502	5514	5527	5539	5551	1	2	4	5	6	7	9	10	11
36	5563	5575	5587	5599	5611	5623	5635	5647	5658	5670	1	2	4	5	6	7	8	10	11
37	5682	5694	5705	5717	5729	5740	5752	5763	5775	5786	1	2	3	5	6	7	8	9	10
38	5798	5809	5821	5832	5843	5855	5866	5877	5888	5899	1	2	3	5	6	7	8	9	10
39	5911	5922	5933	5944	5955	5966	5977	5988	5999	6010	1	2	3	4	5	7	8	9	10
40	6021	6031	6042	6053	6064	6075	6085	6096	6107	6117	1	2	3	4	5	6	8	9	10
41	6128	6138	6149	6160	6170	6180	6191	6201	6212	6222	1	2	3	4	5	6	7	8	9
42	6232	6243	6253	6263	6274	6284	6294	6304	6314	6325	1	2	3	4	5	6	7	8	9
43	6335	6345	6355	6365	6375	6385	6395	6405	6415	6425	1	2	3	4	5	6	7	8	9
44	6435	6444	6454	6464	6474	6484	6493	6503	6513	6522	1	2	3	4	5	6	7	8	9
45	6532	6542	6551	6561	6571	6580	6590	6599	6609	6618	1	2	3	4	5	6	7	8	9
46	6628	6637	6646	6656	6665	6675	6684	6693	6702	6712	1	2	3	4	5	6	7	7	8
47	6721	6730	6739	6749	6758	6767	6776	6785	6794	6803	1	2	3	4	5	5	6	7	8
48	6812	6821	6830	6839	6848	6857	6866	6875	6884	6893	1	2	3	4	4	5	6	7	8
49	6902	6911	6920	6928	6937	6946	6955	6964	6972	6981	1	2	3	4	4	5	6	7	8
50	6990	6998	7007	7016	7024	7033	7042	7050	7059	7067	1	2	3	3	4	5	6	7	8
51	7076	7084	7093	7101	7110	7118	7126	7135	7143	7152	1	2	3	3	4	5	6	7	8
52	7160	7168	7177	7185	7193	7202	7210	7218	7226	7235	1	2	2	3	4	5	6	7	7
53	7243	7251	7259	7267	7275	7284	7292	7300	7308	7316	1	2	2	3	4	5	6	6	7
54	7324	7332	7340	7348	7356	7364	7372	7380	7388	7396	1	2	2	3	4	5	6	6	7

Logarithms

	0	1	2	3	4	5	6	7	8	9	Mean Differences								
											1	2	3	4	5	6	7	8	9
55	7404	7412	7419	7427	7435	7443	7451	7459	7466	7474	1	2	2	3	4	5	5	6	7
56	7482	7490	7497	7505	7513	7520	7528	7536	7543	7551	1	2	2	3	4	5	5	6	7
57	7559	7566	7574	7582	7589	7597	7604	7612	7619	7627	1	2	2	3	4	5	5	6	7
58	7634	7642	7649	7657	7664	7672	7679	7686	7694	7701	1	1	2	3	4	4	5	6	7
59	7709	7716	7723	7731	7738	7745	7752	7760	7767	7774	1	1	2	3	4	4	5	6	7
60	7782	7789	7796	7803	7810	7818	7825	7832	7839	7846	1	1	2	3	4	4	5	6	6
61	7853	7860	7868	7875	7882	7889	7896	7903	7910	7917	1	1	2	3	4	4	5	6	6
62	7924	7931	7938	7945	7952	7959	7966	7973	7980	7987	1	1	2	3	3	4	5	6	6
63	7993	8000	8007	8014	8021	8028	8035	8041	8048	8055	1	1	2	3	3	4	5	5	6
64	8062	8069	8075	8082	8089	8096	8102	8109	8116	8122	1	1	2	3	3	4	5	5	6
65	8129	8136	8142	8149	8156	8162	8169	8176	8182	8189	1	1	2	3	3	4	5	5	6
66	8195	8202	8209	8215	8222	8228	8235	8241	8248	8254	1	1	2	3	3	4	5	5	6
67	8261	8267	8274	8280	8287	8293	8299	8306	8312	8319	1	1	2	3	3	4	5	5	6
68	8325	8331	8338	8344	8351	8357	8363	8370	8376	8382	1	1	2	3	3	4	4	5	6
69	8388	8395	8401	8407	8414	8420	8426	8432	8439	8445	1	1	2	2	3	4	4	5	6
70	8451	8457	8463	8470	8476	8482	8488	8494	8500	8506	1	1	2	2	3	4	4	5	6
71	8513	8519	8525	8531	8537	8543	8549	8555	8561	8567	1	1	2	2	3	4	4	5	5
72	8573	8579	8585	8591	8597	8603	8609	8615	8621	8627	1	1	2	2	3	4	4	5	5
73	8633	8639	8645	8651	8657	8663	8669	8675	8681	8686	1	1	2	2	3	4	4	5	5
74	8692	8698	8704	8710	8716	8722	8727	8733	8739	8745	1	1	2	2	3	4	4	5	5
75	8751	8756	8762	8768	8774	8779	8785	8791	8797	8802	1	1	2	2	3	3	4	5	5
76	8808	8814	8820	8825	8831	8837	8842	8848	8854	8859	1	1	2	2	3	3	4	5	5
77	8865	8871	8876	8882	8887	8893	8899	8904	8910	8915	1	1	2	2	3	3	4	4	5
78	8921	8927	8932	8938	8943	8949	8954	8960	8965	8971	1	1	2	2	3	3	4	4	5
79	8976	8982	8987	8993	8998	9004	9009	9015	9020	9025	1	1	2	2	3	3	4	4	5
80	9031	9036	9042	9047	9053	9058	9063	9069	9074	9079	1	1	2	2	3	3	4	4	5
81	9085	9090	9096	9101	9106	9112	9117	9122	9128	9133	1	1	2	2	3	3	4	4	5
82	9138	9143	9149	9154	9159	9165	9170	9175	9180	9186	1	1	2	2	3	3	4	4	5
83	9191	9196	9201	9206	9212	9217	9222	9227	9232	9238	1	1	2	2	3	3	4	4	5
84	9243	9248	9253	9258	9263	9269	9274	9279	9284	9289	1	1	2	2	3	3	4	4	5
85	9294	9299	9304	9309	9315	9320	9325	9330	9335	9340	1	1	2	2	3	3	4	4	5
86	9345	9350	9355	9360	9365	9370	9375	9380	9385	9390	1	1	2	2	3	3	4	4	5
87	9395	9400	9405	9410	9415	9420	9425	9430	9435	9440	0	1	1	2	2	3	3	4	4
88	9445	9450	9455	9460	9465	9469	9474	9479	9484	9489	0	1	1	2	2	3	3	4	4
89	9494	9499	9504	9509	9513	9518	9523	9528	9533	9538	0	1	1	2	2	3	3	4	4
90	9542	9547	9552	9557	9562	9566	9571	9576	9581	9586	0	1	1	2	2	3	3	4	4
91	9590	9595	9600	9605	9609	9614	9619	9624	9628	9633	0	1	1	2	2	3	3	4	4
92	9638	9643	9647	9653	9657	9661	9666	9671	9675	9680	0	1	1	2	2	3	3	4	4
93	9685	9689	9694	9699	9703	9708	9713	9717	9722	9727	0	1	1	2	2	3	3	4	4
94	9731	9736	9741	9745	9750	9754	9759	9763	9768	9773	0	1	1	2	2	3	3	4	4
95	9777	9782	9786	9791	9795	9800	9805	9809	9814	9818	0	1	1	2	2	3	3	4	4
96	9823	9827	9832	9836	9841	9845	9850	9854	9859	9863	0	1	1	2	2	3	3	4	4
97	9868	9872	9877	9881	9886	9890	9894	9899	9903	9908	0	1	1	2	2	3	3	4	4
98	9912	9917	9921	9926	9930	9934	9939	9943	9948	9952	0	1	1	2	2	3	3	4	4
99	9956	9961	9965	9969	9974	9978	9983	9987	9991	9996	0	1	1	2	2	3	3	3	4

Antilogarithms

	0	1	2	3	4	5	6	7	8	9	1	2	3	4	5	6	7	8	9
											Mean Differences								

	0	**1**	**2**	**3**	**4**	**5**	**6**	**7**	**8**	**9**	**1**	**2**	**3**	**4**	**5**	**6**	**7**	**8**	**9**
·00	1000	1002	1005	1007	1009	1012	1014	1016	1019	1021	0	0	1	1	1	1	2	2	2
·01	1023	1026	1028	1030	1033	1035	1038	1040	1042	1045	0	0	1	1	1	1	2	2	2
·02	1047	1050	1052	1054	1057	1059	1062	1064	1067	1069	0	0	1	1	1	1	2	2	2
·03	1072	1074	1076	1079	1081	1084	1086	1089	1091	1094	0	0	1	1	1	1	2	2	2
·04	1096	1099	1102	1104	1107	1109	1112	1114	1117	1119	0	1	1	1	1	2	2	2	2
·05	1122	1125	1127	1130	1132	1135	1138	1140	1143	1146	0	1	1	1	1	2	2	2	2
·06	1148	1151	1153	1156	1159	1161	1164	1167	1169	1172	0	1	1	1	1	2	2	2	2
·07	1175	1178	1180	1183	1186	1189	1191	1194	1197	1199	0	1	1	1	1	2	2	2	2
·08	1202	1205	1208	1211	1213	1216	1219	1222	1225	1227	0	1	1	1	1	2	2	2	3
·09	1230	1233	1236	1239	1242	1245	1247	1250	1253	1256	0	1	1	1	1	2	2	2	3
·10	1259	1262	1265	1268	1271	1274	1276	1279	1282	1285	0	1	1	1	1	2	2	2	3
·11	1288	1291	1294	1297	1300	1303	1306	1309	1312	1315	0	1	1	1	2	2	2	2	3
·12	1318	1321	1324	1327	1330	1334	1337	1340	1343	1346	0	1	1	1	2	2	2	2	3
·13	1349	1352	1355	1358	1361	1365	1368	1371	1374	1377	0	1	1	1	2	2	2	3	3
·14	1380	1384	1387	1390	1393	1396	1400	1403	1406	1409	0	1	1	1	2	2	2	3	3
·15	1413	1416	1419	1422	1426	1429	1432	1435	1439	1442	0	1	1	1	2	2	2	3	3
·16	1445	1449	1452	1455	1459	1462	1466	1469	1472	1476	0	1	1	1	2	2	2	3	3
·17	1479	1483	1486	1489	1493	1496	1500	1503	1507	1510	0	1	1	1	2	2	2	3	3
·18	1514	1517	1521	1524	1528	1531	1535	1538	1542	1545	0	1	1	1	2	2	2	3	3
·19	1549	1552	1556	1560	1563	1567	1570	1574	1578	1581	0	1	1	1	2	2	3	3	3
·20	1585	1589	1592	1596	1600	1603	1607	1611	1614	1618	0	1	1	1	2	2	3	3	3
·21	1622	1626	1629	1633	1637	1641	1644	1648	1652	1656	0	1	1	2	2	2	3	3	3
·22	1660	1663	1667	1671	1675	1679	1683	1687	1690	1694	0	1	1	2	2	2	3	3	3
·23	1698	1702	1706	1710	1714	1718	1722	1726	1730	1734	0	1	1	2	2	2	3	3	4
·24	1738	1742	1746	1750	1754	1758	1762	1766	1770	1774	0	1	1	2	2	2	3	3	4
·25	1778	1782	1786	1791	1795	1799	1803	1807	1811	1816	0	1	1	2	2	2	3	3	4
·26	1820	1824	1828	1832	1837	1841	1845	1849	1854	1858	0	1	1	2	2	3	3	3	4
·27	1862	1866	1871	1875	1879	1884	1888	1892	1897	1901	0	1	1	2	2	3	3	3	4
·28	1905	1910	1914	1919	1923	1928	1932	1936	1941	1945	0	1	1	2	2	3	3	4	4
·29	1950	1954	1959	1963	1968	1972	1977	1982	1986	1991	0	1	1	2	2	3	3	4	4
·30	1995	2000	2004	2009	2014	2018	2023	2028	2032	2037	0	1	1	2	2	3	3	4	4
·31	2042	2046	2051	2056	2061	2065	2070	2075	2080	2084	0	1	1	2	2	3	3	4	4
·32	2089	2094	2099	2104	2109	2113	2118	2123	2128	2133	0	1	1	2	2	3	3	4	4
·33	2138	2143	2148	2153	2158	2163	2168	2173	2178	2183	0	1	1	2	2	3	3	4	4
·34	2188	2193	2198	2203	2208	2213	2218	2223	2228	2234	1	1	2	2	3	3	4	4	5
·35	2239	2244	2249	2254	2259	2265	2270	2275	2280	2286	1	1	2	2	3	3	4	4	5
·36	2291	2296	2301	2307	2312	2317	2323	2328	2333	2339	1	1	2	2	3	3	4	4	5
·37	2344	2350	2355	2360	2366	2371	2377	2382	2388	2393	1	1	2	2	3	3	4	4	5
·38	2399	2404	2410	2415	2421	2427	2432	2438	2443	2449	1	1	2	2	3	3	4	4	5
·39	2455	2460	2466	2472	2477	2483	2489	2495	2500	2506	1	1	2	2	3	3	4	5	5
·40	2512	2518	2523	2529	2535	2541	2547	2553	2559	2564	1	1	2	2	3	4	4	5	5
·41	2570	2576	2582	2588	2594	2600	2606	2612	2618	2624	1	1	2	2	3	4	4	5	5
·42	2630	2636	2642	2649	2655	2661	2667	2673	2679	2685	1	1	2	2	3	4	4	5	6
·43	2692	2698	2704	2710	2716	2723	2729	2735	2742	2748	1	1	2	3	3	4	4	5	6
·44	2754	2761	2767	2773	2780	2786	2793	2799	2805	2812	1	1	2	3	3	4	4	5	6
·45	2818	2825	2831	2838	2844	2851	2858	2864	2871	2877	1	1	2	3	3	4	5	5	6
·46	2884	2891	2897	2904	2911	2917	2924	2931	2938	2944	1	1	2	3	3	4	5	5	6
·47	2951	2958	2965	2972	2979	2985	2992	2999	3006	3013	1	1	2	3	3	4	5	5	6
·48	3020	3027	3034	3041	3048	3055	3062	3069	3076	3083	1	1	2	3	4	4	5	6	6
·49	3090	3097	3105	3112	3119	3126	3133	3141	3148	3155	1	1	2	3	4	4	5	6	6

Antilogarithms

	0	1	2	3	4	5	6	7	8	9	1	2	3	4	5	6	7	8	9
														Mean Differences					
·50	3162	3170	3177	3184	3192	3199	3206	3214	3221	3228	1	1	2	3	4	4	5	6	7
·51	3236	3243	3251	3258	3266	3273	3281	3289	3296	3304	1	2	2	3	4	5	5	6	7
·52	3311	3319	3327	3334	3342	3350	3357	3365	3373	3381	1	2	2	3	4	5	5	6	7
·53	3388	3396	3404	3412	3420	3428	3436	3443	3451	3459	1	2	2	3	4	5	6	6	7
·54	3467	3475	3483	3491	3499	3508	3516	3524	3532	3540	1	2	2	3	4	5	6	6	7
·55	3548	3556	3565	3573	3581	3589	3597	3606	3614	3622	1	2	2	3	4	5	6	7	7
·56	3631	3639	3648	3656	3664	3673	3681	3690	3698	3707	1	2	3	3	4	5	6	7	8
·57	3715	3724	3733	3741	3750	3758	3767	3776	3784	3793	1	2	3	3	4	5	6	7	8
·58	3802	3811	3819	3828	3837	3846	3855	3864	3873	3882	1	2	3	4	4	5	6	7	8
·59	3890	3899	3908	3917	3926	3936	3945	3954	3963	3972	1	2	3	4	5	5	6	7	8
·60	3981	3990	3999	4009	4018	4027	4036	4046	4055	4064	1	2	3	4	5	6	6	7	8
·61	4074	4083	4093	4102	4111	4121	4130	4140	4150	4159	1	2	3	4	5	6	7	8	9
·62	4169	4178	4188	4198	4207	4217	4227	4236	4246	4256	1	2	3	4	5	6	7	8	9
·63	4266	4276	4285	4295	4305	4315	4325	4335	4345	4355	1	2	3	4	5	6	7	8	9
·64	4365	4375	4385	4395	4406	4416	4426	4436	4446	4457	1	2	3	4	5	6	7	8	9
·65	4467	4477	4487	4498	4508	4519	4529	4539	4550	4560	1	2	3	4	5	6	7	8	9
·66	4571	4581	4592	4603	4613	4624	4634	4645	4656	4667	1	2	3	4	5	6	7	9	10
·67	4677	4688	4699	4710	4721	4732	4742	4753	4764	4775	1	2	3	4	5	7	8	9	10
·68	4786	4797	4808	4819	4831	4842	4853	4864	4875	4887	1	2	3	4	6	7	8	9	10
·69	4898	4909	4920	4932	4943	4955	4966	4977	4989	5000	1	2	3	5	6	7	8	9	10
·70	5012	5023	5035	5047	5058	5070	5082	5093	5105	5117	1	2	4	5	6	7	8	9	11
·71	5129	5140	5152	5164	5176	5188	5200	5212	5224	5236	1	2	4	5	6	7	8	10	11
·72	5248	5260	5272	5284	5297	5309	5321	5333	5346	5358	1	2	4	5	6	7	9	10	11
·73	5370	5383	5395	5408	5420	5433	5445	5458	5470	5483	1	3	4	5	6	8	9	10	11
·74	5495	5508	5521	5534	5546	5559	5572	5585	5598	5610	1	3	4	5	6	8	9	10	12
·75	5623	5636	5649	5662	5675	5689	5702	5715	5728	5741	1	3	4	5	7	8	9	10	12
·76	5754	5768	5781	5794	5808	5821	5834	5848	5861	5875	1	3	4	5	7	8	9	11	12
·77	5888	5902	5916	5929	5943	5957	5970	5984	5998	6012	1	3	4	5	7	8	10	11	12
·78	6026	6039	6053	6067	6081	6095	6109	6124	6138	6152	1	3	4	6	7	8	10	11	13
·79	6166	6180	6194	6209	6223	6237	6252	6266	6281	6295	1	3	4	6	7	9	10	11	13
·80	6310	6324	6339	6353	6368	6383	6397	6412	6427	6442	1	3	4	6	7	9	10	12	13
·81	6457	6471	6486	6501	6516	6531	6546	6561	6577	6592	2	3	5	6	8	9	11	12	14
·82	6607	6622	6637	6653	6668	6683	6699	6714	6730	6745	2	3	5	6	8	9	11	13	14
·83	6761	6776	6792	6808	6823	6839	6855	6871	6887	6902	2	3	5	6	8	9	11	13	14
·84	6918	6934	6950	6966	6982	6998	7015	7031	7047	7063	2	3	5	6	8	10	11	13	15
·85	7079	7096	7112	7129	7145	7161	7178	7194	7211	7228	2	3	5	7	8	10	12	13	15
·86	7244	7261	7278	7295	7311	7328	7345	7362	7379	7396	2	3	5	7	8	10	12	13	15
·87	7413	7430	7447	7464	7482	7499	7516	7534	7551	7568	2	3	5	7	9	10	12	14	16
·88	7586	7603	7621	7638	7656	7674	7691	7709	7727	7745	2	4	5	7	9	11	12	14	16
·89	7762	7780	7798	7816	7834	7852	7870	7889	7907	7925	2	4	5	7	9	11	13	14	16
·90	7943	7962	7980	7998	8017	8035	8054	8072	8091	8110	2	4	6	7	9	11	13	15	17
·91	8128	8147	8166	8185	8204	8222	8241	8260	8279	8299	2	4	6	8	9	11	13	15	17
·92	8318	8337	8356	8375	8395	8414	8433	8453	8472	8492	2	4	6	8	10	12	14	15	17
·93	8511	8531	8551	8570	8590	8610	8630	8650	8670	8690	2	4	6	8	10	12	14	16	18
·94	8710	8730	8750	8770	8790	8810	8831	8851	8872	8892	2	4	6	8	10	12	14	16	18
·95	8913	8933	8954	8974	8995	9016	9036	9057	9078	9099	2	4	6	8	10	12	15	17	19
·96	9120	9141	9162	9183	9204	9226	9247	9268	9290	9311	2	4	6	8	11	13	15	17	19
·97	9333	9354	9376	9397	9419	9441	9462	9484	9506	9528	2	4	7	9	11	13	15	17	20
·98	9550	9572	9594	9616	9638	9661	9683	9705	9727	9750	2	4	7	9	11	13	16	18	20
·99	9772	9795	9817	9840	9863	9886	9908	9931	9954	9977	2	5	7	9	11	14	16	18	20

	0' 0·0°	6' 0·1°	12' 0·2°	18' 0·3°	24' 0·4°	30' 0·5°	36' 0·6°	42' 0·7°	48' 0·8°	54' 0·9°	Mean Differences 1'	2'	3'	4'	5'
0°	0·0000	0017	0035	0052	0070	0087	0105	0122	0140	0157	3	6	9	12	15
1	0·0175	0192	0209	0227	0244	0262	0279	0297	0314	0332	3	6	9	12	15
2	0·0349	0366	0384	0401	0419	0436	0454	0471	0488	0506	3	6	9	12	15
3	0·0523	0541	0558	0576	0593	0610	0628	0645	0663	0680	3	6	9	12	15
4	0·0698	0715	0732	0750	0767	0785	0802	0819	0837	0854	3	6	9	12	15
5	0·0872	0889	0906	0924	0941	0958	0976	0993	1011	1028	3	6	9	12	14
6	0·1045	1063	1080	1097	1115	1132	1149	1167	1184	1201	3	6	9	12	14
7	0·1219	1236	1253	1271	1288	1305	1323	1340	1357	1374	3	6	9	12	14
8	0·1392	1409	1426	1444	1461	1478	1495	1513	1530	1547	3	6	9	12	14
9	0·1564	1582	1599	1616	1633	1650	1668	1685	1702	1719	3	6	9	12	14
10°	0·1736	1754	1771	1788	1805	1822	1840	1857	1874	1891	3	6	9	11	14
11	0·1908	1925	1942	1959	1977	1994	2011	2028	2045	2062	3	6	9	11	14
12	0·2079	2096	2113	2130	2147	2164	2181	2198	2215	2233	3	6	9	11	14
13	0·2250	2267	2284	2300	2317	2334	2351	2368	2385	2402	3	6	8	11	14
14	0·2419	2436	2453	2470	2487	2504	2521	2538	2554	2571	3	6	8	11	14
15	0·2588	2605	2622	2639	2656	2672	2689	2706	2723	2740	3	6	8	11	14
16	0·2756	2773	2790	2807	2823	2840	2857	2874	2890	2907	3	6	8	11	14
17	0·2924	2940	2957	2974	2990	3007	3024	3040	3057	3074	3	6	8	11	14
18	0·3090	3107	3123	3140	3156	3173	3190	3206	3223	3239	3	6	8	11	14
19	0·3256	3272	3289	3305	3322	3338	3355	3371	3387	3404	3	5	8	11	14
20°	0·3420	3437	3453	3469	3486	3502	3518	3535	3551	3567	3	5	8	11	14
21	0·3584	3600	3616	3633	3649	3665	3681	3697	3714	3730	3	5	8	11	14
22	0·3746	3762	3778	3795	3811	3827	3843	3859	3875	3891	3	5	8	11	14
23	0·3907	3923	3939	3955	3971	3987	4003	4019	4035	4051	3	5	8	11	14
24	0·4067	4083	4099	4115	4131	4147	4163	4179	4195	4210	3	5	8	11	13
25	0·4226	4242	4258	4274	4289	4305	4321	4337	4352	4368	3	5	8	11	13
26	0·4384	4399	4415	4431	4446	4462	4478	4493	4509	4524	3	5	8	10	13
27	0·4540	4555	4571	4586	4602	4617	4633	4648	4664	4679	3	5	8	10	13
28	0·4695	4710	4726	4741	4756	4772	4787	4802	4818	4833	3	5	8	10	13
29	0·4848	4863	4879	4894	4909	4924	4939	4955	4970	4985	3	5	8	10	13
30°	0·5000	5015	5030	5045	5060	5075	5090	5105	5120	5135	3	5	8	10	13
31	0·5150	5165	5180	5195	5210	5225	5240	5255	5270	5284	2	5	7	10	12
32	0·5299	5314	5329	5344	5358	5373	5388	5402	5417	5432	2	5	7	10	12
33	0·5446	5461	5476	5490	5505	5519	5534	5548	5563	5577	2	5	7	10	12
34	0·5592	5606	5621	5635	5650	5664	5678	5693	5707	5721	2	5	7	10	12
35	0·5736	5750	5764	5779	5793	5807	5821	5835	5850	5864	2	5	7	9	12
36	0·5878	5892	5906	5920	5934	5948	5962	5976	5990	6004	2	5	7	9	12
37	0·6018	6032	6046	6060	6074	6088	6101	6115	6129	6143	2	5	7	9	12
38	0·6157	6170	6184	6198	6211	6225	6239	6252	6266	6280	2	5	7	9	11
39	0·6293	6307	6320	6334	6347	6361	6374	6388	6401	6414	2	4	7	9	11
40°	0·6428	6441	6455	6468	6481	6494	6508	6521	6534	6547	2	4	7	9	11
41	0·6561	6574	6587	6600	6613	6626	6639	6652	6665	6678	2	4	7	9	11
42	0·6691	6704	6717	6730	6743	6756	6769	6782	6794	6807	2	4	6	9	11
43	0·6820	6833	6845	6858	6871	6884	6896	6909	6921	6934	2	4	6	8	11
44	0·6947	6959	6972	6984	6997	7009	7022	7034	7046	7059	2	4	6	8	10

Natural Sines

	0' 0·0°	6' 0·1°	12' 0·2°	18' 0·3°	24' 0·4°	30' 0·5°	36' 0·6°	42' 0·7°	48' 0·8°	54' 0·9°	Mean Differences 1' 2' 3'			4' 5'	
45°	0·7071	7083	7096	7108	7120	7133	7145	7157	7169	7181	2	4	6	8	10
46	0·7193	7206	7218	7230	7242	7254	7266	7278	7290	7302	2	4	6	8	10
47	0·7314	7325	7337	7349	7361	7373	7385	7396	7408	7420	2	4	6	8	10
48	0·7431	7443	7455	7466	7478	7490	7501	7513	7524	7536	2	4	6	8	10
49	0·7547	7559	7570	7581	7593	7604	7615	7627	7638	7649	2	4	6	8	9
50°	0·7660	7672	7683	7694	7705	7716	7727	7738	7749	7760	2	4	6	7	9
51	0·7771	7782	7793	7804	7815	7826	7837	7848	7859	7869	2	4	5	7	9
52	0·7880	7891	7902	7912	7923	7934	7944	7955	7965	7976	2	4	5	7	9
53	0·7986	7997	8007	8018	8028	8039	8049	8059	8070	8080	2	3	5	7	9
54	0·8090	8100	8111	8121	8131	8141	8151	8161	8171	8181	2	3	5	7	8
55	0·8192	8202	8211	8221	8231	8241	8251	8261	8271	8281	2	3	5	7	8
56	0·8290	8300	8310	8320	8329	8339	8348	8358	8368	8377	2	3	5	6	8
57	0·8387	8396	8406	8415	8425	8434	8443	8453	8462	8471	2	3	5	6	8
58	0·8480	8490	8499	8508	8517	8526	8536	8545	8554	8563	2	3	5	6	8
59	0·8572	8581	8590	8599	8607	8616	8625	8634	8643	8652	1	3	4	6	7
60°	0·8660	8669	8678	8686	8695	8704	8712	8721	8729	8738	1	3	4	6	7
61	0·8746	8755	8763	8771	8780	8788	8796	8805	8813	8821	1	3	4	6	7
62	0·8829	8838	8846	8854	8862	8870	8878	8886	8894	8902	1	3	4	5	7
63	0·8910	8918	8926	8934	8942	8949	8957	8965	8973	8980	1	3	4	5	6
64	0·8988	8996	9003	9011	9018	9026	9033	9041	9048	9056	1	3	4	5	6
65	0·9063	9070	9078	9085	9092	9100	9107	9114	9121	9128	1	2	4	5	6
66	0·9135	9143	9150	9157	9164	9171	9178	9184	9191	9198	1	2	3	5	6
67	0·9205	9212	9219	9225	9232	9239	9245	9252	9259	9265	1	2	3	4	6
68	0·9272	9278	9285	9291	9298	9304	9311	9317	9323	9330	1	2	3	4	5
69	0·9336	9342	9348	9354	9361	9367	9373	9379	9385	9391	1	2	3	4	5
70°	0·9397	9403	9409	9415	9421	9426	9432	9438	9444	9449	1	2	3	4	5
71	0·9455	9461	9466	9472	9478	9483	9489	9494	9500	9505	1	2	3	4	5
72	0·9511	9516	9521	9527	9532	9537	9542	9548	9553	9558	1	2	3	3	4
73	0·9563	9568	9573	9578	9583	9588	9593	9598	9603	9608	1	2	2	3	4
74	0·9613	9617	9622	9627	9632	9636	9641	9646	9650	9655	1	2	2	3	4
75	0·9659	9664	9668	9673	9677	9681	9686	9690	9694	9699	1	1	2	3	4
76	0·9703	9707	9711	9715	9720	9724	9728	9732	9736	9740	1	1	2	3	3
77	0·9744	9748	9751	9755	9759	9763	9767	9770	9774	9778	1	1	2	3	3
78	0·9781	9785	9789	9792	9796	9799	9803	9806	9810	9813	1	1	2	2	3
79	0·9816	9820	9823	9826	9829	9833	9836	9839	9842	9845	1	1	2	2	3
80°	0·9848	9851	9854	9857	9860	9863	9866	9869	9871	9874	0	1	1	2	2
81	0·9877	9880	9882	9885	9888	9890	9893	9895	9898	9900	0	1	1	2	2
82	0·9903	9905	9907	9910	9912	9914	9917	9919	9921	9923	0	1	1	2	2
83	0·9925	9928	9930	9932	9934	9936	9938	9940	9942	9943	0	1	1	1	2
84	0·9945	9947	9949	9951	9952	9954	9956	9957	9959	9960	0	1	1	1	2
85	0·9962	9963	9965	9966	9968	9969	9971	9972	9973	9974	0	0	1	1	1
86	0·9976	9977	9978	9979	9980	9981	9982	9983	9984	9985	0	0	1	1	1
87	0·9986	9987	9988	9989	9990	9990	9991	9992	9993	9993	0	0	0	1	1
88	0·9994	9995	9995	9996	9996	9997	9997	9997	9998	9998	0	0	0	0	0
89	0·9998	9999	9999	9999	9999	1·000	1·000	1·000	1·000	1·000	0	0	0	0	0

Natural Cosines

	0' 0·0°	6' 0·1°	12' 0·2°	18' 0·3°	24' 0·4°	30' 0·5°	36' 0·6°	42' 0·7°	48' 0·8°	54' 0·9°	1'	2'	3'	4'	5'
0°	1·000	1·000	1·000	1·000	1·000	1·000	·9999	9999	9999	9999	0	0	0	0	0
1	0·9998	9998	9998	9997	9997	9997	9996	9996	9995	9995	0	0	0	0	0
2	0·9994	9993	9993	9992	9991	9990	9990	9989	9988	9987	0	0	0	1	1
3	0·9986	9985	9984	9983	9982	9981	9980	9979	9978	9977	0	0	1	1	1
4	0·9976	9974	9973	9972	9971	9969	9968	9966	9965	9963	0	0	1	1	1
5	0·9962	9960	9959	9957	9956	9954	9952	9951	9949	9947	0	1	1	1	2
6	0·9945	9943	9942	9940	9938	9936	9934	9932	9930	9928	0	1	1	1	2
7	0·9925	9923	9921	9919	9917	9914	9912	9910	9907	9905	0	1	1	2	2
8	0·9903	9900	9898	9895	9893	9890	9888	9885	9882	9880	0	1	1	2	2
9	0·9877	9874	9871	9869	9866	9863	9860	9857	9854	9851	0	1	1	2	2
10°	0·9848	9845	9842	9839	9836	9833	9829	9826	9823	9820	1	1	2	2	3
11	0·9816	9813	9810	9806	9803	9799	9796	9792	9789	9785	1	1	2	2	3
12	0·9781	9778	9774	9770	9767	9763	9759	9755	9751	9748	1	1	2	3	3
13	0·9744	9740	9736	9732	9728	9724	9720	9715	9711	9707	1	1	2	3	3
14	0·9703	9699	9694	9690	9686	9681	9677	9673	9668	9664	1	1	2	3	4
15	0·9659	9655	9650	9646	9641	9636	9632	9627	9622	9617	1	2	2	3	4
16	0·9613	9608	9603	9598	9593	9588	9583	9578	9573	9568	1	2	2	3	4
17	0·9563	9558	9553	9548	9542	9537	9532	9527	9521	9516	1	2	3	3	4
18	0·9511	9505	9500	9494	9489	9483	9478	9472	9466	9461	1	2	3	4	5
19	0·9455	9449	9444	9438	9432	9426	9421	9415	9409	9403	1	2	3	4	5
20°	0·9397	9391	9385	9379	9373	9367	9361	9354	9348	9342	1	2	3	4	5
21	0·9336	9330	9323	9317	9311	9304	9298	9291	9285	9278	1	2	3	4	5
22	0·9272	9265	9259	9252	9245	9239	9232	9225	9219	9212	1	2	3	4	6
23	0·9205	9198	9191	9184	9178	9171	9164	9157	9150	9143	1	2	3	5	6
24	0·9135	9128	9121	9114	9107	9100	9092	9085	9078	9070	1	2	4	5	6
25	0·9063	9056	9048	9041	9033	9026	9018	9011	9003	8996	1	3	4	5	6
26	0·8988	8980	8973	8965	8957	8949	8942	8934	8926	8918	1	3	4	5	6
27	0·8910	8902	8894	8886	8878	8870	8862	8854	8846	8838	1	3	4	5	7
28	0·8829	8821	8813	8805	8796	8788	8780	8771	8763	8755	1	3	4	6	7
29	0·8746	8738	8729	8721	8712	8704	8695	8686	8678	8669	1	3	4	6	7
30°	0·8660	8652	8643	8634	8625	8616	8607	8599	8590	8581	1	3	4	6	7
31	0·8572	8563	8554	8545	8536	8526	8517	8508	8499	8490	2	3	5	6	8
32	0·8480	8471	8462	8453	8443	8434	8425	8415	8406	8396	2	3	5	6	8
33	0·8387	8377	8368	8358	8348	8339	8329	8320	8310	8300	2	3	5	6	8
34	0·8290	8281	8271	8261	8251	8241	8231	8221	8211	8202	2	3	5	7	8
35	0·8192	8181	8171	8161	8151	8141	8131	8121	8111	8100	2	3	5	7	8
36	0·8090	8080	8070	8059	8049	8039	8028	8018	8007	7997	2	3	5	7	9
37	0·7986	7976	7965	7955	7944	7934	7923	7912	7902	7891	2	4	5	7	9
38	0·7880	7869	7859	7848	7837	7826	7815	7804	7793	7782	2	4	5	7	9
39	0·7771	7760	7749	7738	7727	7716	7705	7694	7683	7672	2	4	6	7	9
40°	0·7660	7649	7638	7627	7615	7604	7593	7581	7570	7559	2	4	6	8	9
41	0·7547	7536	7524	7513	7501	7490	7478	7466	7455	7443	2	4	6	8	10
42	0·7431	7420	7408	7396	7385	7373	7361	7349	7337	7325	2	4	6	8	10
43	0·7314	7302	7290	7278	7266	7254	7242	7230	7218	7206	2	4	6	8	10
44	0·7193	7181	7169	7157	7145	7133	7120	7108	7096	7083	2	4	6	8	10

SUBTRACT Mean Differences

	0' 0·0°	6' 0·1°	12' 0·2°	18' 0·3°	24' 0·4°	30' 0·5°	36' 0·6°	42' 0·7°	48' 0·8°	54' 0·9°	SUBTRACT Mean Differences				
											1'	2'	3'	4'	5'
45°	0·7071	7059	7046	7034	7022	7009	6997	6984	6972	6959	2	4	6	8	10
46	0·6947	6934	6921	6909	6896	6884	6871	6858	6845	6833	2	4	6	8	11
47	0·6820	6807	6794	6782	6769	6756	6743	6730	6717	6704	2	4	6	9	11
48	0·6691	6678	6665	6652	6639	6626	6613	6600	6587	6574	2	4	7	9	11
49	0·6561	6547	6534	6521	6508	6494	6481	6468	6455	6441	2	4	7	9	11
50°	0·6428	6414	6401	6388	6374	6361	6347	6334	6320	6307	2	4	7	9	11
51	0·6293	6280	6266	6252	6239	6225	6211	6198	6184	6170	2	5	7	9	11
52	0·6157	6143	6129	6115	6101	6088	6074	6060	6046	6032	2	5	7	9	12
53	0·6018	6004	5990	5976	5962	5948	5934	5920	5906	5892	2	5	7	9	12
54	0·5878	5864	5850	5835	5821	5807	5793	5779	5764	5750	2	5	7	9	12
55	0·5736	5721	5707	5693	5678	5664	5650	5635	5621	5606	2	5	7	10	12
56	0·5592	5577	5563	5548	5534	5519	5505	5490	5476	5461	2	5	7	10	12
57	0·5446	5432	5417	5402	5388	5373	5358	5344	5329	5314	2	5	7	10	12
58	0·5299	5284	5270	5255	5240	5225	5210	5195	5180	5165	2	5	7	10	12
59	0·5150	5135	5120	5105	5090	5075	5060	5045	5030	5015	3	5	8	10	13
60°	0·5000	4985	4970	4955	4939	4924	4909	4894	4879	4863	3	5	8	10	13
61	0·4848	4833	4818	4802	4787	4772	4756	4741	4726	4710	3	5	8	10	13
62	0·4695	4679	4664	4648	4633	4617	4602	4586	4571	4555	3	5	8	10	13
63	0·4540	4524	4509	4493	4478	4462	4446	4431	4415	4399	3	5	8	11	13
64	0·4384	4368	4352	4337	4321	4305	4289	4274	4258	4242	3	5	8	11	13
65	0·4226	4210	4195	4179	4163	4147	4131	4115	4099	4083	3	5	8	11	13
66	0·4067	4051	4035	4019	4003	3987	3971	3955	3939	3923	3	5	8	11	14
67	0·3907	3891	3875	3859	3843	3827	3811	3795	3778	3762	3	5	8	11	14
68	0·3746	3730	3714	3697	3681	3665	3649	3633	3616	3600	3	5	8	11	14
69	0·3584	3567	3551	3535	3518	3502	3486	3469	3453	3437	3	5	8	11	14
70°	0·3420	3404	3387	3371	3355	3338	3322	3305	3289	3272	3	5	8	11	14
71	0·3256	3239	3223	3206	3190	3173	3156	3140	3123	3107	3	6	8	11	14
72	0·3090	3074	3057	3040	3024	3007	2990	2974	2957	2940	3	6	8	11	14
73	0·2924	2907	2890	2874	2857	2840	2823	2807	2790	2773	3	6	8	11	14
74	0·2756	2740	2723	2706	2689	2672	2656	2639	2622	2605	3	6	8	11	14
75	0·2588	2571	2554	2538	2521	2504	2487	2470	2453	2436	3	6	8	11	14
76	0·2419	2402	2385	2368	2351	2334	2317	2300	2284	2267	3	6	8	11	14
77	0·2250	2233	2215	2198	2181	2164	2147	2130	2113	2096	3	6	9	11	14
78	0·2079	2062	2045	2028	2011	1994	1977	1959	1942	1925	3	6	9	11	14
79	0·1908	1891	1874	1857	1840	1822	1805	1788	1771	1754	3	6	9	11	14
80°	0·1736	1719	1702	1685	1668	1650	1633	1616	1599	1582	3	6	9	12	14
81	0·1564	1547	1530	1513	1495	1478	1461	1444	1426	1409	3	6	9	12	14
82	0·1392	1374	1357	1340	1323	1305	1288	1271	1253	1236	3	6	9	12	14
83	0·1219	1201	1184	1167	1149	1132	1115	1097	1080	1063	3	6	9	12	14
84	0·1045	1028	1011	0993	0976	0958	0941	0924	0906	0889	3	6	9	12	14
85	0·0872	0854	0837	0819	0802	0785	0767	0750	0732	0715	3	6	9	12	14
86	0·0698	0680	0663	0645	0628	0610	0593	0576	0558	0541	3	6	9	12	15
87	0·0523	0506	0488	0471	0454	0436	0419	0401	0384	0366	3	6	9	12	15
88	0·0349	0332	0314	0297	0279	0262	0244	0227	0209	0192	3	6	9	12	15
89	0·0175	0157	0140	0122	0105	0087	0070	0052	0035	0017	3	6	9	12	15

	0' 0·0°	6' 0·1°	12' 0·2°	18' 0·3°	24' 0·4°	30' 0·5°	36' 0·6°	42' 0·7°	48' 0·8°	54' 0·9°	Mean Differences 1' 2' 3'		4' 5'
0°	0·0000	0017	0035	0052	0070	0087	0105	0122	0140	0157	3 6 9		12 15
1	0·0175	0192	0209	0227	0244	0262	0279	0297	0314	0332	3 6 9		12 15
2	0·0349	0367	0384	0402	0419	0437	0454	0472	0489	0507	3 6 9		12 15
3	0·0524	0542	0559	0577	0594	0612	0629	0647	0664	0682	3 6 9		12 15
4	0·0699	0717	0734	0752	0769	0787	0805	0822	0840	0857	3 6 9		12 15
5	0·0875	0892	0910	0928	0945	0963	0981	0998	1016	1033	3 6 9		12 15
6	0·1051	1069	1086	1104	1122	1139	1157	1175	1192	1210	3 6 9		12 15
7	0·1228	1246	1263	1281	1299	1317	1334	1352	1370	1388	3 6 9		12 15
8	0·1405	1423	1441	1459	1477	1495	1512	1530	1548	1566	3 6 9		12 15
9	0·1584	1602	1620	1638	1655	1673	1691	1709	1727	1745	3 6 9		12 15
10°	0·1763	1781	1799	1817	1835	1853	1871	1890	1908	1926	3 6 9		12 15
11	0·1944	1962	1980	1998	2016	2035	2053	2071	2089	2107	3 6 9		12 15
12	0·2126	2144	2162	2180	2199	2217	2235	2254	2272	2290	3 6 9		12 15
13	0·2309	2327	2345	2364	2382	2401	2419	2438	2456	2475	3 6 9		12 15
14	0·2493	2512	2530	2549	2568	2586	2605	2623	2642	2661	3 6 9		13 16
15	0·2679	2698	2717	2736	2754	2773	2792	2811	2830	2849	3 6 9		13 16
16	0·2867	2886	2905	2924	2943	2962	2981	3000	3019	3038	3 6 9		13 16
17	0·3057	3076	3096	3115	3134	3153	3172	3191	3211	3230	3 6 10		13 16
18	0·3249	3269	3288	3307	3327	3346	3365	3385	3404	3424	3 6 10		13 16
19	0·3443	3463	3482	3502	3522	3541	3561	3581	3600	3620	3 7 10		13 16
20°	0·3640	3659	3679	3699	3719	3739	3759	3779	3799	3819	3 7 10		13 17
21	0·3839	3859	3879	3899	3919	3939	3959	3979	4000	4020	3 7 10		13 17
22	0·4040	4061	4081	4101	4122	4142	4163	4183	4204	4224	3 7 10		14 17
23	0·4245	4265	4286	4307	4327	4348	4369	4390	4411	4431	3 7 10		14 17
24	0·4452	4473	4494	4515	4536	4557	4578	4599	4621	4642	4 7 11		14 18
25	0·4663	4684	4706	4727	4748	4770	4791	4813	4834	4856	4 7 11		14 18
26	0·4877	4899	4921	4942	4964	4986	5008	5029	5051	5073	4 7 11		15 18
27	0·5095	5117	5139	5161	5184	5206	5228	5250	5272	5295	4 7 11		15 18
28	0·5317	5340	5362	5384	5407	5430	5452	5475	5498	5520	4 8 11		15 19
29	0·5543	5566	5589	5612	5635	5658	5681	5704	5727	5750	4 8 12		15 19
30°	0·5774	5797	5820	5844	5867	5890	5914	5938	5961	5985	4 8 12		16 20
31	0·6009	6032	6056	6080	6104	6128	6152	6176	6200	6224	4 8 12		16 20
32	0·6249	6273	6297	6322	6346	6371	6395	6420	6445	6469	4 8 12		16 20
33	0·6494	6519	6544	6569	6594	6619	6644	6669	6694	6720	4 8 13		17 21
34	0·6745	6771	6796	6822	6847	6873	6899	6924	6950	6976	4 9 13		17 21
35	0·7002	7028	7054	7080	7107	7133	7159	7186	7212	7239	4 9 13		18 22
36	0·7265	7292	7319	7346	7373	7400	7427	7454	7481	7508	5 9 14		18 23
37	0·7536	7563	7590	7618	7646	7673	7701	7729	7757	7785	5 9 14		18 23
38	0·7813	7841	7869	7898	7926	7954	7983	8012	8040	8069	5 9 14		19 24
39	0·8098	8127	8156	8185	8214	8243	8273	8302	8332	8361	5 10 15		20 24
40°	0·8391	8421	8451	8481	8511	8541	8571	8601	8632	8662	5 10 15		20 25
41	0·8693	8724	8754	8785	8816	8847	8878	8910	8941	8972	5 10 16		21 26
42	0·9004	9036	9067	9099	9131	9163	9195	9228	9260	9293	5 11 16		21 27
43	0·9325	9358	9391	9424	9457	9490	9523	9556	9590	9623	6 11 17		22 28
44	0·9657	9691	9725	9759	9793	9827	9861	9896	9930	9965	6 11 17		23 29

Natural Cotangents:

Cot x° = tan (90—x)° and use above table.

	0' 0·0°	6' 0·1°	12' 0·2°	18' 0·3°	24' 0·4°	30' 0·5°	36' 0·6°	42' 0·7°	48' 0·8°	54' 0·9°	Mean Differences 1'	2'	3'	4'	5'
45°	1·0000	0035	0070	0105	0141	0176	0212	0247	0283	0319	6	12	18	24	30
46	1·0355	0392	0428	0464	0501	0538	0575	0612	0649	0686	6	12	18	25	31
47	1·0724	0761	0799	0837	0875	0913	0951	0990	1028	1067	6	13	19	25	32
48	1·1106	1145	1184	1224	1263	1303	1343	1383	1423	1463	7	13	20	27	33
49	1·1504	1544	1585	1626	1667	1708	1750	1792	1833	1875	7	14	21	28	34
50°	1·1918	1960	2002	2045	2088	2131	2174	2218	2261	2305	7	14	22	29	36
51	1·2349	2393	2437	2484	2527	2572	2617	2662	2708	2753	8	15	23	30	38
52	1·2799	2846	2892	2938	2985	3032	3079	3127	3175	3222	8	16	24	31	39
53	1·3270	3319	3367	3416	3465	3514	3564	3613	3663	3713	8	16	25	33	41
54	1·3764	3814	3865	3916	3968	4019	4071	4124	4176	4229	9	17	26	34	43
55	1·4281	4335	4388	4442	4496	4550	4605	4659	4715	4770	9	18	27	36	45
56	1·4826	4882	4938	4994	5051	5108	5166	5224	5282	5340	10	19	29	38	48
57	1·5399	5458	5517	5577	5637	5697	5757	5818	5880	5941	10	20	30	40	50
58	1·6003	6066	6128	6191	6255	6319	6383	6447	6512	6577	10	21	32	43	53
59	1·6643	6709	6775	6842	6909	6977	7045	7113	7182	7251	11	23	34	45	56
60°	1·7321	7391	7461	7532	7603	7675	7747	7820	7893	7966	12	24	36	48	60
61	1·8040	8115	8190	8265	8341	8418	8495	8572	8650	8728	13	26	38	51	64
62	1·8807	8887	8967	9047	9128	9210	9292	9375	9458	9542	14	27	41	55	68
63	1·9626	9711	9797	9883	9970	0057	0145	0233	0323	0413	15	29	44	58	73
64	2·0503	0594	0686	0778	0872	0965	1060	1155	1251	1348	16	31	47	63	78
65	2·1445	1543	1642	1742	1842	1943	2045	2148	2251	2355	17	34	51	68	85
66	2·2460	2566	2673	2781	2889	2998	3109	3220	3332	3445	18	37	55	73	92
67	2·3559	3673	3789	3906	4023	4142	4262	4383	4504	4627	20	40	60	79	99
68	2·4751	4876	5002	5129	5257	5386	5517	5649	5782	5916	22	43	65	87	108
69	2·6051	6187	6325	6464	6605	6746	6889	7034	7179	7326	24	47	71	95	119
70°	2·7475	7625	7776	7929	8083	8239	8397	8556	8716	8878	26	52	78	104	131
71	2·9042	9208	9375	9544	9714	9887	0061	0237	0415	0595	29	58	87	116	145
72	3·0777	0961	1146	1334	1524	1716	1910	2106	2305	2506	32	64	96	129	161
73	3·2709	2914	3122	3332	3544	3759	3977	4197	4420	4646	36	72	108	144	180
74	3·4874	5105	5339	5576	5816	6059	6305	6554	6806	7062	41	81	122	163	204
75	3·7321	7583	7848	8118	8391	8667	8947	9232	9520	9812	46	93	139	186	232
76	4·0108	0408	0713	1022	1335	1653	1976	2303	2635	2972					
77	4·3315	3662	4015	4374	4737	5107	5483	5864	6252	6646					
78	4·7046	7453	7867	8288	8716	9152	9594	0045	0504	0970					
79	5·1446	1929	2422	2924	3435	3955	4486	5026	5578	6140					
80°	5·6713	7297	7894	8502	9124	9758	0405	1066	1742	2432					
81	6·3138	3859	4596	5350	6122	6912	7720	8548	9395	0264	*Mean* differences				
82	7·1154	2066	3002	3962	4947	5958	6996	8062	9158	0285	no longer				
83	8·1443	2636	3863	5126	6427	7769	9152	0579	2052	3572	sufficiently				
84	9·514	9·677	9·845	10·02	10·20	10·39	10·58	10·78	10·99	11·20	accurate.				
85	11·43	11·66	11·91	12·16	12·43	12·71	13·00	13·30	13·62	13·95					
86	14·30	14·67	15·06	15·46	15·89	16·35	16·83	17·34	17·89	18·46					
87	19·08	19·74	20·45	21·20	22·02	22·90	23·86	24·90	26·03	27·27					
88	28·64	30·14	31·82	33·69	35·80	38·19	40·92	44·07	47·74	52·08					
89	57·29	63·66	71·62	81·85	95·49	114·6	143·2	191·0	286·5	573·0					

Natural Cotangents:

Cot x° = tan (90—x)° and use above table.

Natural Secants

	0' 0·0°	6' 0·1°	12' 0·2°	18' 0·3°	24' 0·4°	30' 0·5°	36' 0·6°	42' 0·7°	48' 0·8°	54' 0·9°	Mean Differences				
											1'	2'	3'	4'	5'
0°	1·0000	0000	0000	0000	0000	0000	0001	0001	0001	0001	0	0	0	0	0
1	1·0002	0002	0002	0003	0003	0003	0004	0004	0005	0006	0	0	0	0	0
2	1·0006	0007	0007	0008	0009	0010	0010	0011	0012	0013	0	0	0	1	1
3	1·0014	0015	0016	0017	0018	0019	0020	0021	0022	0023	0	0	1	1	1
4	1·0024	0026	0027	0028	0030	0031	0032	0034	0035	0037	0	0	1	1	1
5	1·0038	0040	0041	0043	0045	0046	0048	0050	0051	0053	0	1	1	1	2
6	1·0055	0057	0059	0061	0063	0065	0067	0069	0071	0073	0	1	1	1	2
7	1·0075	0077	0079	0082	0084	0086	0089	0091	0093	0096	0	1	1	2	2
8	1·0098	0101	0103	0106	0108	0111	0114	0116	0119	0122	0	1	1	2	2
9	1·0125	0127	0130	0133	0136	0139	0142	0145	0148	0151	0	1	1	2	2
10°	1·0154	0157	0161	0164	0167	0170	0174	0177	0180	0184	1	1	2	2	3
11	1·0187	0191	0194	0198	0201	0205	0209	0212	0216	0220	1	1	2	2	3
12	1·0223	0227	0231	0235	0239	0243	0247	0251	0255	0259	1	1	2	3	3
13	1·0263	0267	0271	0276	0280	0284	0288	0293	0297	0302	1	1	2	3	4
14	1·0306	0311	0315	0320	0324	0329	0334	0338	0343	0348	1	2	2	3	4
15	1·0353	0358	0363	0367	0372	0377	0382	0388	0393	0398	1	2	3	3	4
16	1·0403	0408	0413	0419	0424	0429	0435	0440	0446	0451	1	2	3	4	4
17	1·0457	0463	0468	0474	0480	0485	0491	0497	0503	0509	1	2	3	4	5
18	1·0515	0521	0527	0533	0539	0545	0551	0557	0564	0570	1	2	3	4	5
19	1·0576	0583	0589	0595	0602	0608	0615	0622	0628	0635	1	2	3	4	5
20°	1·0642	0649	0655	0662	0669	0676	0683	0690	0697	0704	1	2	3	5	6
21	1·0711	0719	0726	0733	0740	0748	0755	0763	0770	0778	1	2	4	5	6
22	1·0785	0793	0801	0808	0816	0824	0832	0840	0848	0856	1	3	4	5	7
23	1·0864	0872	0880	0888	0896	0904	0913	0921	0929	0938	1	3	4	5	7
24	1·0946	0955	0963	0972	0981	0989	0998	1007	1016	1025	1	3	4	6	7
25	1·1034	1043	1052	1061	1070	1079	1089	1098	1107	1117	2	3	5	6	8
26	1·1126	1136	1145	1155	1164	1174	1184	1194	1203	1213	2	3	5	6	8
27	1·1223	1233	1243	1253	1264	1274	1284	1294	1305	1315	2	3	5	7	9
28	1·1326	1336	1347	1357	1368	1379	1390	1401	1412	1423	2	4	5	7	9
29	1·1434	1445	1456	1467	1478	1490	1501	1512	1524	1535	2	4	6	8	9
30°	1·1547	1559	1570	1582	1594	1606	1618	1630	1642	1654	2	4	6	8	10
31	1·1666	1679	1691	1703	1716	1728	1741	1753	1766	1779	2	4	6	8	10
32	1·1792	1805	1818	1831	1844	1857	1870	1883	1897	1910	2	4	7	9	11
33	1·1924	1937	1951	1964	1978	1992	2006	2020	2034	2048	2	5	7	9	11
34	1·2062	2076	2091	2105	2120	2134	2149	2163	2178	2193	2	5	7	10	12
35	1·2208	2223	2238	2253	2268	2283	2299	2314	2329	2345	3	5	8	10	13
36	1·2361	2376	2392	2408	2424	2440	2456	2472	2489	2505	2	5	8	11	13
37	1·2521	2538	2554	2571	2588	2605	2622	2639	2656	2673	3	6	8	11	14
38	1·2690	2708	2725	2742	2760	2778	2796	2813	2831	2849	3	6	9	12	15
39	1·2868	2886	2904	2923	2941	2960	2978	2997	3016	3035	3	6	9	12	15
40°	1·3054	3073	3093	3112	3131	3151	3171	3190	3210	3230	3	7	10	13	16
41	1·3250	3270	3291	3311	3331	3352	3373	3393	3414	3435	3	7	10	14	17
42	1·3456	3478	3499	3520	3542	3563	3585	3607	3629	3651	4	7	11	14	18
43	1·3673	3696	3718	3741	3763	3786	3809	3832	3855	3878	4	8	11	15	19
44	1·3902	3925	3949	3972	3996	4020	4044	4069	4093	4118	4	8	12	16	20

Natural Cosecants:

Cosec x° = sec (90—x)° and use above table.

	0' 0·0°	6' 0·1°	12' 0·2°	18' 0·3°	24' 0·4°	30' 0·5°	36' 0·6°	42' 0·7°	48' 0·8°	54' 0·9°	Mean Differences				
											1'	2'	3'	4'	5'
45°	1·4142	4167	4192	4217	4242	4267	4293	4318	4344	4370	4	8	13	17	21
46	1·4396	4422	4448	4474	4501	4527	4554	4581	4608	4635	4	9	13	18	22
47	1·4663	4690	4718	4746	4774	4802	4830	4859	4887	4916	5	9	14	19	23
48	1·4945	4974	5003	5032	5062	5092	5121	5151	5182	5212	5	10	15	20	25
49	1·5243	5273	5304	5335	5366	5398	5429	5461	5493	5525	5	10	16	21	26
50°	1·5557	5590	5622	5655	5688	5721	5755	5788	5822	5856	6	11	17	22	28
51	1·5890	5925	5959	5994	6029	6064	6099	6135	6171	6207	6	12	18	24	29
52	1·6243	6279	6316	6353	6390	6427	6464	6502	6540	6578	6	12	19	25	31
53	1·6616	6655	6694	6733	6772	6812	6852	6892	6932	6972	7	13	20	26	33
54	1·7013	7054	7095	7137	7179	7221	7263	7305	7348	7391	7	14	21	28	35
55	1·7434	7478	7522	7566	7610	7655	7700	7745	7791	7837	7	15	22	30	37
56	1·7883	7929	7976	8023	8070	8118	8166	8214	8263	8312	8	16	24	32	40
57	1·8361	8410	8460	8510	8561	8612	8663	8714	8766	8818	8	17	25	34	42
58	1·8871	8924	8977	9031	9084	9139	9194	9249	9304	9360	9	18	27	36	45
59	1·9416	9473	9530	9587	9645	9703	9762	9821	9880	9940	10	19	29	39	49
60°	2·0000	0061	0122	0183	0245	0308	0371	0434	0498	0562	10	21	31	42	52
61	2·0627	0692	0757	0824	0890	0957	1025	1093	1162	1231	11	22	34	45	56
62	2·1301	1371	1441	1513	1584	1657	1730	1803	1877	1952	11	24	36	48	60
63	2·2027	2103	2179	2256	2333	2412	2490	2570	2650	2730	13	26	39	52	65
64	2·2812	2894	2976	3060	3144	3228	3314	3400	3486	3574	14	28	42	57	71
65	2·3662	3751	3841	3931	4022	4114	4207	4300	4395	4490	15	31	46	62	77
66	2·4586	4683	4780	4879	4978	5078	5180	5282	5384	5488					
67	2·5593	5699	5805	5913	6022	6131	6242	6354	6466	6580					
68	2·6695	6811	6927	7046	7165	7285	7407	7529	7653	7778					
69	2·7904	8032	8161	8291	8422	8555	8688	8824	8960	9099					
70°	2·9238	9379	9521	9665	9811	9957	0106	0256	0407	0561					
71	3·0716	0872	1030	1190	1352	1515	1681	1848	2017	2188					
72	3·2361	2535	2712	2891	3072	3255	3440	3628	3817	4009					
73	3·4203	4399	4598	4799	5003	5209	5418	5629	5843	6060					
74	3·6280	6502	6727	6955	7186	7420	7657	7897	8140	8387					
75	3·8637	8890	9147	9408	9672	9939	0211	0486	0765	1048					
76	4·1336	1627	1923	2223	2527	2837	3150	3469	3792	4121					
77	4·4454	4793	5137	5486	5841	6202	6569	6942	7321	7706					
78	4·8097	8496	8901	9313	9732	0159	0593	1034	1484	1942					
79	5·2408	2883	3367	3860	4362	4874	5396	5928	6470	7023					
80°	5·7588	8164	8751	9351	9963	0589	1227	1880	2546	3228					
81	6·3925	4637	5366	6111	6874	7655	8454	9273	0112	0972					
82	7·1853	2757	3684	4635	5611	6613	7642	8700	9787	0905					
83	8·2055	3238	4457	5711	7004	8337	9711	1129	2593	4105					
84	9·5668	7283	8955	0685	2477	4334	6261	8260	0336	2493					
85	11·474	11·71	11·95	12·20	12·47	12·75	13·03	13·34	13·65	13·99					
86	14·34	14·70	15·09	15·50	15·93	16·38	16·86	17·37	17·91	18·49					
87	19·11	19·77	20·47	21·23	22·04	22·93	23·88	24·92	26·05	27·29					
88	28·65	30·16	31·84	33·71	35·81	38·20	40·93	44·08	47·75	52·09					
89	57·30	63·66	71·62	81·85	95·49	114·6	143·2	191·0	286·5	573·0					

Mean differences no longer sufficiently accurate.

Natural Cosecants:
Cosec x° = sec (90—x)° and use above table.

Logarithmic Sines

	0' 0·0°	6' 0·1°	12' 0·2°	18' 0·3°	24' 0·4°	30' 0·5°	36' 0·6°	42' 0·7°	48' 0·8°	54' 0·9°	Mean Differences 1'	2'	3'	4'	5'
0°	$-\infty$	$\bar{3}$·2419	5429	7190	8439	9408	$\bar{0}$200	$\bar{0}$870	$\bar{1}$450	$\bar{1}$961					
1	$\bar{2}$·2419	2832	3210	3558	3880	4179	4459	4723	4971	5206					
2	$\bar{2}$·5428	5640	5842	6035	6220	6397	6567	6731	6889	7041					
3	$\bar{2}$·7188	7330	7468	7602	7731	7857	7979	8098	8213	8326					
4	$\bar{2}$·8436	8543	8647	8749	8849	8946	9042	9135	9226	9315	16	32	48	64	80
5	$\bar{2}$·9403	9489	9573	9655	9736	9816	9894	9970	$\bar{0}$046	$\bar{0}$120	13	26	39	52	65
6	$\bar{1}$·0192	0264	0334	0403	0472	0539	0605	0670	0734	0797	11	22	33	44	55
7	$\bar{1}$·0859	0920	0981	1040	1099	1157	1214	1271	1326	1381	10	19	29	38	48
8	$\bar{1}$·1436	1489	1542	1594	1646	1697	1747	1797	1847	1895	8	17	25	34	42
9	$\bar{1}$·1943	1991	2038	2085	2131	2176	2221	2266	2310	2353	8	15	23	30	38
10°	$\bar{1}$·2397	2439	2482	2524	2565	2606	2647	2687	2727	2767	7	14	20	27	34
11	$\bar{1}$·2806	2845	2883	2921	2959	2997	3034	3070	3107	3143	6	12	19	25	31
12	$\bar{1}$·3179	3214	3250	3284	3319	3353	3387	3421	3455	3488	6	11	17	23	28
13	$\bar{1}$·3521	3554	3586	3618	3650	3682	3713	3745	3775	3806	5	11	16	21	26
14	$\bar{1}$·3837	3867	3897	3927	3957	3986	4015	4044	4073	4102	5	10	15	20	24
15	$\bar{1}$·4130	4158	4186	4214	4242	4269	4296	4323	4350	4377	5	9	14	18	23
16	$\bar{1}$·4403	4430	4456	4482	4508	4533	4559	4584	4609	4634	4	9	13	17	21
17	$\bar{1}$·4659	4684	4709	4733	4757	4781	4805	4829	4853	4876	4	8	12	16	20
18	$\bar{1}$·4900	4923	4946	4969	4992	5015	5037	5060	5082	5104	4	8	11	15	19
19	$\bar{1}$·5126	5148	5170	5192	5213	5235	5256	5278	5299	5320	4	7	11	14	18
20°	$\bar{1}$·5341	5361	5382	5402	5423	5443	5463	5484	5504	5523	3	7	10	14	17
21	$\bar{1}$·5543	5563	5583	5602	5621	5641	5660	5679	5698	5717	3	6	10	13	16
22	$\bar{1}$·5736	5754	5773	5792	5810	5828	5847	5865	5883	5901	3	6	9	12	15
23	$\bar{1}$·5919	5937	5954	5972	5990	6007	6024	6042	6059	6076	3	6	9	12	15
24	$\bar{1}$·6093	6110	6127	6144	6161	6177	6194	6210	6227	6243	3	6	8	11	14
25	$\bar{1}$·6259	6276	6292	6308	6324	6340	6356	6371	6387	6403	3	5	8	11	13
26	$\bar{1}$·6418	6434	6449	6465	6480	6495	6510	6526	6541	6556	3	5	8	10	13
27	$\bar{1}$·6570	6585	6600	6615	6629	6644	6659	6673	6687	6702	2	5	7	10	12
28	$\bar{1}$·6716	6730	6744	6759	6773	6787	6801	6814	6828	6842	2	5	7	9	12
29	$\bar{1}$·6856	6869	6883	6896	6910	6923	6937	6950	6963	6977	2	4	7	9	11
30°	$\bar{1}$·6990	7003	7016	7029	7042	7055	7068	7080	7093	7106	2	4	6	9	11
31	$\bar{1}$·7118	7131	7144	7156	7168	7181	7193	7205	7218	7230	2	4	6	8	10
32	$\bar{1}$·7242	7254	7266	7278	7290	7302	7314	7326	7338	7349	2	4	6	8	10
33	$\bar{1}$·7361	7373	7384	7396	7407	7419	7430	7442	7453	7464	2	4	6	8	10
34	$\bar{1}$·7476	7487	7498	7509	7520	7531	7542	7553	7564	7575	2	4	6	7	9
35	$\bar{1}$·7586	7597	7607	7618	7629	7640	7650	7661	7671	7682	2	4	5	7	9
36	$\bar{1}$·7692	7703	7713	7723	7734	7744	7754	7764	7774	7785	2	3	5	7	9
37	$\bar{1}$·7795	7805	7815	7825	7835	7844	7854	7864	7874	7884	2	3	5	7	8
38	$\bar{1}$·7893	7903	7913	7922	7932	7941	7951	7960	7970	7979	2	3	5	6	8
39	$\bar{1}$·7989	7998	8007	8017	8026	8035	8044	8053	8063	8072	2	3	5	6	8
40°	$\bar{1}$·8081	8090	8099	8108	8117	8125	8134	8143	8152	8161	1	3	4	6	7
41	$\bar{1}$·8169	8178	8187	8195	8204	8213	8221	8230	8238	8247	1	3	4	6	7
42	$\bar{1}$·8255	8264	8272	8280	8289	8297	8305	8313	8322	8330	1	3	4	6	7
43	$\bar{1}$·8338	8346	8354	8362	8370	8378	8386	8394	8402	8410	1	3	4	5	7
44	$\bar{1}$·8418	8426	8433	8441	8449	8457	8464	8472	8480	8487	1	3	4	5	6

	0' 0·0°	6' 0·1°	12' 0·2°	18' 0·3°	24' 0·4°	30' 0·5°	36' 0·6°	42' 0·7°	48' 0·8°	54' 0·9°	1'	2'	3'	4'	5'
45°	$\bar{1}$·8495	8502	8510	8517	8525	8532	8540	8547	8555	8562	1	2	4	5	6
46	$\bar{1}$·8569	8577	8584	8591	8598	8606	8613	8620	8627	8634	1	2	4	5	6
47	$\bar{1}$·8641	8648	8655	8662	8669	8676	8683	8690	8697	8704	1	2	3	5	6
48	$\bar{1}$·8711	8718	8724	8731	8738	8745	8751	8758	8765	8771	1	2	3	4	6
49	$\bar{1}$·8778	8784	8791	8797	8804	8810	8817	8823	8830	8836	1	2	3	4	5
50°	$\bar{1}$·8843	8849	8855	8862	8868	8874	8880	8887	8893	8899	1	2	3	4	5
51	$\bar{1}$·8905	8911	8917	8923	8929	8935	8941	8947	8953	8959	1	2	3	4	5
52	$\bar{1}$·8965	8971	8977	8983	8989	8995	9000	9006	9012	9018	1	2	3	4	5
53	$\bar{1}$·9023	9029	9035	9041	9046	9052	9057	9063	9069	9074	1	2	3	4	5
54	$\bar{1}$·9080	9085	9091	9096	9101	9107	9112	9118	9123	9128	1	2	3	4	5
55	$\bar{1}$·9134	9139	9144	9149	9155	9160	9165	9170	9175	9181	1	2	3	3	4
56	$\bar{1}$·9186	9191	9196	9201	9206	9211	9216	9221	9226	9231	1	2	3	3	4
57	$\bar{1}$·9236	9241	9246	9251	9255	9260	9265	9270	9275	9279	1	2	2	3	4
58	$\bar{1}$·9284	9289	9294	9298	9303	9308	9312	9317	9322	9326	1	2	2	3	4
59	$\bar{1}$·9331	9335	9340	9344	9349	9353	9358	9362	9367	9371	1	1	2	3	4
60°	$\bar{1}$·9375	9380	9384	9388	9393	9397	9401	9406	9410	9414	1	1	2	3	4
61	$\bar{1}$·9418	9422	9427	9431	9435	9439	9443	9447	9451	9455	1	1	2	3	3
62	$\bar{1}$·9459	9463	9467	9471	9475	9479	9483	9487	9491	9495	1	1	2	3	3
63	$\bar{1}$·9499	9503	9507	9510	9514	9518	9522	9525	9529	9533	1	1	2	3	3
64	$\bar{1}$·9537	9540	9544	9548	9551	9555	9558	9562	9566	9569	1	1	2	2	3
65	$\bar{1}$·9573	9576	9580	9583	9587	9590	9594	9597	9601	9604	1	1	2	2	3
66	$\bar{1}$·9607	9611	9614	9617	9621	9624	9627	9631	9634	9637	1	1	2	2	3
67	$\bar{1}$·9640	9643	9647	9650	9653	9656	9659	9662	9666	9669	1	1	2	2	3
68	$\bar{1}$·9672	9675	9678	9681	9684	9687	9690	9693	9696	9699	0	1	1	2	2
69	$\bar{1}$·9702	9704	9707	9710	9713	9716	9719	9722	9724	9727	0	1	1	2	2
70°	$\bar{1}$·9730	9733	9735	9738	9741	9743	9746	9749	9751	9754	0	1	1	2	2
71	$\bar{1}$·9757	9759	9762	9764	9767	9770	9772	9775	9777	9780	0	1	1	2	2
72	$\bar{1}$·9782	9785	9787	9789	9792	9794	9797	9799	9801	9804	0	1	1	2	2
73	$\bar{1}$·9806	9808	9811	9813	9815	9817	9820	9822	9824	9826	0	1	1	2	2
74	$\bar{1}$·9828	9831	9833	9835	9837	9839	9841	9843	9845	9847	0	1	1	1	2
75	$\bar{1}$·9849	9851	9853	9855	9857	9859	9861	9863	9865	9867	0	1	1	1	2
76	$\bar{1}$·9869	9871	9873	9875	9876	9878	9880	9882	9884	9885	0	1	1	1	2
77	$\bar{1}$·9887	9889	9891	9892	9894	9896	9897	9899	9901	9902	0	1	1	1	1
78	$\bar{1}$·9904	9906	9907	9909	9910	9912	9913	9915	9916	9918	0	1	1	1	1
79	$\bar{1}$·9919	9921	9922	9924	9925	9927	9928	9929	9931	9932	0	0	1	1	1
80°	$\bar{1}$·9934	9935	9936	9937	9939	9940	9941	9943	9944	9945	0	0	1	1	1
81	$\bar{1}$·9946	9947	9949	9950	9951	9952	9953	9954	9955	9956	0	0	1	1	1
82	$\bar{1}$·9958	9959	9960	9961	9962	9963	9964	9965	9966	9967	0	0	1	1	1
83	$\bar{1}$·9968	9968	9969	9970	9971	9972	9973	9974	9975	9975	0	0	1	1	1
84	$\bar{1}$·9976	9977	9978	9978	9979	9980	9981	9981	9982	9983	0	0	0	0	1
85	$\bar{1}$·9983	9984	9985	9985	9986	9987	9987	9988	9988	9989	0	0	0	0	0
86	$\bar{1}$·9989	9990	9990	9991	9991	9992	9992	9993	9993	9994	0	0	0	0	0
87	$\bar{1}$·9994	9994	9995	9995	9996	9996	9996	9996	9997	9997	0	0	0	0	0
88	$\bar{1}$·9997	9998	9998	9998	9998	9999	9999	9999	9999	9999	0	0	0	0	0
89	$\bar{1}$·9999	9999	$\bar{0}$000	$\bar{0}$000	$\bar{0}$000	$\bar{0}$000	$\bar{0}$000	$\bar{0}$000	$\bar{0}$000	$\bar{0}$000	0	0	0	0	0

Mean Differences — columns 1', 2', 3', 4', 5'

	0' 0·0°	6' 0·1°	12' 0·2°	18' 0·3°	24' 0·4°	30' 0·5°	36' 0·6°	42' 0·7°	48' 0·8°	54' 0·9°	SUBTRACT Mean Differences				
											1'	2'	3'	4'	5'
0°	0·0000	0000	0000	0000	0000	0000	0000	0000	0000	$\bar{1}$·9999	0	0	0	0	0
1	$\bar{1}$·9999	9999	9999	9999	9999	9999	9998	9998	9998	9998	0	0	0	0	0
2	$\bar{1}$·9997	9997	9997	9996	9996	9996	9996	9995	9995	9994	0	0	0	0	0
3	$\bar{1}$·9994	9994	9993	9993	9992	9992	9991	9991	9990	9990	0	0	0	0	0
4	$\bar{1}$·9989	9989	9988	9988	9987	9987	9986	9985	9985	9984	0	0	0	0	0
5	$\bar{1}$·9983	9983	9982	9981	9981	9980	9979	9978	9978	9977	0	0	0	0	1
6	$\bar{1}$·9976	9975	9975	9974	9973	9972	9971	9970	9969	9968	0	0	0	1	1
7	$\bar{1}$·9968	9967	9966	9965	9964	9963	9962	9961	9960	9959	0	0	1	1	1
8	$\bar{1}$·9958	9956	9955	9954	9953	9952	9951	9950	9949	9947	0	0	1	1	1
9	$\bar{1}$·9946	9945	9944	9943	9941	9940	9939	9937	9936	9935	0	0	1	1	1
10°	$\bar{1}$·9934	9932	9931	9929	9928	9927	9925	9924	9922	9921	0	0	1	1	1
11	$\bar{1}$·9919	9918	9916	9915	9913	9912	9910	9909	9907	9906	0	1	1	1	1
12	$\bar{1}$·9904	9902	9901	9899	9897	9896	9894	9892	9891	9889	0	1	1	1	1
13	$\bar{1}$·9887	9885	9884	9882	9880	9878	9876	9875	9873	9871	0	1	1	1	2
14	$\bar{1}$·9869	9867	9865	9863	9861	9859	9857	9855	9853	9851	0	1	1	1	2
15	$\bar{1}$·9849	9847	9845	9843	9841	9839	9837	9835	9833	9831	0	1	1	1	2
16	$\bar{1}$·9828	9826	9824	9822	9820	9817	9815	9813	9811	9808	0	1	1	2	2
17	$\bar{1}$·9806	9804	9801	9799	9797	9794	9792	9789	9787	9785	0	1	1	2	2
18	$\bar{1}$·9782	9780	9777	9775	9772	9770	9767	9764	9762	9759	0	1	1	2	2
19	$\bar{1}$·9757	9754	9751	9749	9746	9743	9741	9738	9735	9733	0	1	1	2	2
20°	$\bar{1}$·9730	9727	9724	9722	9719	9716	9713	9710	9707	9704	0	1	1	2	2
21	$\bar{1}$·9702	9699	9696	9693	9690	9687	9684	9681	9678	9675	0	1	1	2	2
22	$\bar{1}$·9672	9669	9666	9662	9659	9656	9653	9650	9647	9643	1	1	2	2	3
23	$\bar{1}$·9640	9637	9634	9631	9627	9624	9621	9617	9614	9611	1	1	2	2	3
24	$\bar{1}$·9607	9604	9601	9597	9594	9590	9587	9583	9580	9576	1	1	2	2	3
25	$\bar{1}$·9573	9569	9566	9562	9558	9555	9551	9548	9544	9540	1	1	2	2	3
26	$\bar{1}$·9537	9533	9529	9525	9522	9518	9514	9510	9507	9503	1	1	2	3	3
27	$\bar{1}$·9499	9495	9491	9487	9483	9479	9475	9471	9467	9463	1	1	2	3	3
28	$\bar{1}$·9459	9455	9451	9447	9443	9439	9435	9431	9427	9422	1	1	2	3	3
29	$\bar{1}$·9418	9414	9410	9406	9401	9397	9393	9388	9384	9380	1	1	2	3	4
30°	$\bar{1}$·9375	9371	9367	9362	9358	9353	9349	9344	9340	9335	1	1	2	3	4
31	$\bar{1}$·9331	9326	9322	9317	9312	9308	9303	9298	9294	9289	1	2	2	3	4
32	$\bar{1}$·9284	9279	9275	9270	9265	9260	9255	9251	9246	9241	1	2	2	3	4
33	$\bar{1}$·9236	9231	9226	9221	9216	9211	9206	9201	9196	9191	1	2	3	3	4
34	$\bar{1}$·9186	9181	9175	9170	9165	9160	9155	9149	9144	9139	1	2	3	3	4
35	$\bar{1}$·9134	9128	9123	9118	9112	9107	9101	9096	9091	9085	1	2	3	4	5
36	$\bar{1}$·9080	9074	9069	9063	9057	9052	9046	9041	9035	9029	1	2	3	4	5
37	$\bar{1}$·9023	9018	9012	9006	9000	8995	8989	8983	8977	8971	1	2	3	4	5
38	$\bar{1}$·8965	8959	8953	8947	8941	8935	8929	8923	8917	8911	1	2	3	4	5
39	$\bar{1}$·8905	8899	8893	8887	8880	8874	8868	8862	8855	8849	1	2	3	4	5
40°	$\bar{1}$·8843	8836	8830	8823	8817	8810	8804	8797	8791	8784	1	2	3	4	5
41	$\bar{1}$·8778	8771	8765	8758	8751	8745	8738	8731	8724	8718	1	2	3	5	6
42	$\bar{1}$·8711	8704	8697	8690	8683	8676	8669	8662	8655	8648	1	2	3	5	6
43	$\bar{1}$·8641	8634	8627	8620	8613	8606	8598	8591	8584	8577	1	2	4	5	6
44	$\bar{1}$·8569	8562	8555	8547	8540	8532	8525	8517	8510	8502	1	2	4	5	6

	0' 0·0°	6' 0·1°	12' 0·2°	18' 0·3°	24' 0·4°	30' 0·5°	36' 0·6°	42' 0·7°	48' 0·8°	54' 0·9°	SUBTRACT Mean Differences				
											1'	2'	3'	4'	5'
45°	1̄·8495	8487	8480	8472	8464	8457	8449	8441	8433	8426	1	3	4	5	6
46	1̄·8418	8410	8402	8394	8386	8378	8370	8362	8354	8346	1	3	4	5	7
47	1̄·8338	8330	8322	8313	8305	8297	8289	8280	8272	8264	1	3	4	6	7
48	1̄·8255	8247	8238	8230	8221	8213	8204	8195	8187	8178	1	3	4	6	7
49	1̄·8169	8161	8152	8143	8134	8125	8117	8108	8099	8090	1	3	4	6	7
50°	1̄·8081	8072	8063	8053	8044	8035	8026	8017	8007	7998	2	3	5	6	8
51	1̄·7989	7979	7970	7960	7951	7941	7932	7922	7913	7903	2	3	5	6	8
52	1̄·7893	7884	7874	7864	7854	7844	7835	7825	7815	7805	2	3	5	7	8
53	1̄·7795	7785	7774	7764	7754	7744	7734	7723	7713	7703	2	3	5	7	9
54	1̄·7692	7682	7671	7661	7650	7640	7629	7618	7607	7597	2	4	5	7	9
55	1̄·7586	7575	7564	7553	7542	7531	7520	7509	7498	7487	2	4	6	7	9
56	1̄·7476	7464	7453	7442	7430	7419	7407	7396	7384	7373	2	4	6	8	10
57	1̄·7361	7349	7338	7326	7314	7302	7290	7278	7266	7254	2	4	6	8	10
58	1̄·7242	7230	7218	7205	7193	7181	7168	7156	7144	7131	2	4	6	8	10
59	1̄·7118	7106	7093	7080	7068	7055	7042	7029	7016	7003	2	4	6	9	11
60°	1̄·6990	6977	6963	6950	6937	6923	6910	6896	6883	6869	2	4	7	9	11
61	1̄·6856	6842	6828	6814	6801	6787	6773	6759	6744	6730	2	5	7	9	12
62	1̄·6716	6702	6687	6673	6659	6644	6629	6615	6600	6585	2	5	7	10	12
63	1̄·6570	6556	6541	6526	6510	6495	6480	6465	6449	6434	3	5	8	10	13
64	1̄·6418	6403	6387	6371	6356	6340	6324	6308	6292	6276	3	5	8	11	13
65	1̄·6259	6243	6227	6210	6194	6177	6161	6144	6127	6110	3	6	8	11	14
66	1̄·6093	6076	6059	6042	6024	6007	5990	5972	5954	5937	3	6	9	12	15
67	1̄·5919	5901	5883	5865	5847	5828	5810	5792	5773	5754	3	6	9	12	15
68	1̄·5736	5717	5698	5679	5660	5641	5621	5602	5583	5563	3	6	10	13	16
69	1̄·5543	5523	5504	5484	5463	5443	5423	5402	5382	5361	3	7	10	14	17
70°	1̄·5341	5320	5299	5278	5256	5235	5213	5192	5170	5148	4	7	11	14	18
71	1̄·5126	5104	5082	5060	5037	5015	4992	4969	4946	4923	4	8	11	15	19
72	1̄·4900	4876	4853	4829	4805	4781	4757	4733	4709	4684	4	8	12	16	20
73	1̄·4659	4634	4609	4584	4559	4533	4508	4482	4456	4430	4	9	13	17	21
74	1̄·4403	4377	4350	4323	4296	4269	4242	4214	4186	4158	5	9	14	18	23
75	1̄·4130	4102	4073	4044	4015	3986	3957	3927	3897	3867	5	10	15	20	24
76	1̄·3837	3806	3775	3745	3713	3682	3650	3618	3586	3554	5	11	16	21	26
77	1̄·3521	3488	3455	3421	3387	3353	3319	3284	3250	3214	6	11	17	23	28
78	1̄·3179	3143	3107	3070	3034	2997	2959	2921	2883	2845	6	12	19	25	31
79	1̄·2806	2767	2727	2687	2647	2606	2565	2524	2482	2439	7	14	20	27	34
80°	1̄·2397	2353	2310	2266	2221	2176	2131	2085	2038	1991	8	15	23	30	38
81	1̄·1943	1895	1847	1797	1747	1697	1646	1594	1542	1489	8	17	25	34	42
82	1̄·1436	1381	1326	1271	1214	1157	1099	1040	0981	0920	10	19	29	38	48
83	1̄·0859	0797	0734	0670	0605	0539	0472	0403	0334	0264	11	22	33	44	55
84	1̄·0192	0120	0046	9̄970	9̄894	9̄816	9̄736	9̄655	9̄573	9̄489	13	26	39	52	65
85	2̄·9403	9315	9226	9135	9042	8946	8849	8749	8647	8543	16	32	48	64	80
86	2̄·8436	8326	8213	8098	7979	7857	7731	7602	7468	7330					
87	2̄·7188	7041	6889	6731	6567	6397	6220	6035	5842	5640					
88	2̄·5428	5206	4971	4723	4459	4179	3880	3558	3210	2832					
89	2̄·2419	1961	1450	0870	0200	9̄408	8̄439	7190	5̄429	2̄419					

	0' 0·0°	6' 0·1°	12' 0·2°	18' 0·3°	24' 0·4°	30' 0·5°	36' 0·6°	42' 0·7°	48' 0·8°	54' 0·9°	Mean Differences 1' 2' 3'	4' 5'
0°	—∞ $\bar{3}$·2419	5429	7190	8439	9409	$\bar{0}$200	$\bar{0}$870	$\bar{1}$450	$\bar{1}$962			
1	$\bar{2}$·2419	2833	3211	3559	3881	4181	4461	4725	4973	5208		
2	$\bar{2}$·5431	5643	5845	6038	6223	6401	6571	6736	6894	7046		
3	$\bar{2}$·7194	7337	7475	7609	7739	7865	7988	8107	8223	8336	16 32 48	64 81
4	$\bar{2}$·8446	8554	8659	8762	8862	8960	9056	9150	9241	9331	13 26 40	53 66
5	$\bar{2}$·9420	9506	9591	9674	9756	9836	9915	9992	$\bar{0}$068	$\bar{0}$143	13 26 40	53 66
6	$\bar{1}$·0216	0289	0360	0430	0499	0567	0633	0699	0764	0828	11 22 34	45 56
7	$\bar{1}$·0891	0954	1015	1076	1135	1194	1252	1310	1367	1423	10 20 29	39 49
8	$\bar{1}$·1478	1533	1587	1640	1693	1745	1797	1848	1898	1948	9 17 26	35 43
9	$\bar{1}$·1997	2046	2094	2142	2189	2236	2282	2328	2374	2419	8 16 23	31 39
10°	$\bar{1}$·2463	2507	2551	2594	2637	2680	2722	2764	2805	2846	7 14 21	28 35
11	$\bar{1}$·2887	2927	2967	3006	3046	3085	3123	3162	3200	3237	6 13 19	26 32
12	$\bar{1}$·3275	3312	3349	3385	3422	3458	3493	3529	3564	3599	6 12 18	24 30
13	$\bar{1}$·3634	3668	3702	3736	3770	3804	3837	3870	3903	3935	6 11 17	22 28
14	$\bar{1}$·3968	4000	4032	4064	4095	4127	4158	4189	4220	4250	5 10 16	21 26
15	$\bar{1}$·4281	4311	4341	4371	4400	4430	4459	4488	4517	4546	5 10 15	20 25
16	$\bar{1}$·4575	4603	4632	4660	4688	4716	4744	4771	4799	4826	5 9 14	19 23
17	$\bar{1}$·4853	4880	4907	4934	4961	4987	5014	5040	5066	5092	4 9 13	18 22
18	$\bar{1}$·5118	5143	5169	5195	5220	5245	5270	5295	5320	5345	4 8 13	17 21
19	$\bar{1}$·5370	5394	5419	5443	5467	5491	5516	5539	5563	5587	4 8 12	16 20
20°	$\bar{1}$·5611	5634	5658	5681	5704	5727	5750	5773	5796	5819	4 8 12	15 19
21	$\bar{1}$·5842	5864	5887	5909	5932	5954	5976	5998	6020	6042	4 7 11	15 19
22	$\bar{1}$·6064	6086	6108	6129	6151	6172	6194	6215	6236	6257	4 7 11	14 18
23	$\bar{1}$·6279	6300	6321	6341	6362	6383	6404	6424	6445	6465	3 7 10	14 17
24	$\bar{1}$·6486	6506	6527	6547	6567	6587	6607	6627	6647	6667	3 7 10	13 17
25	$\bar{1}$·6687	6706	6726	6746	6765	6785	6804	6824	6843	6863	3 7 10	13 16
26	$\bar{1}$·6882	6901	6920	6939	6958	6977	6996	7015	7034	7053	3 6 9	13 16
27	$\bar{1}$·7072	7090	7109	7128	7146	7165	7183	7202	7220	7238	3 6 9	12 15
28	$\bar{1}$·7257	7275	7293	7311	7330	7348	7366	7384	7402	7420	3 6 9	12 15
29	$\bar{1}$·7438	7455	7473	7491	7509	7526	7544	7562	7579	7597	3 6 9	12 15
30°	$\bar{1}$·7614	7632	7649	7667	7684	7701	7719	7736	7753	7771	3 6 9	12 14
31	$\bar{1}$·7788	7805	7822	7839	7856	7873	7890	7907	7924	7941	3 6 9	11 14
32	$\bar{1}$·7958	7975	7992	8008	8025	8042	8059	8075	8092	8109	3 6 8	11 14
33	$\bar{1}$·8125	8142	8158	8175	8191	8208	8224	8241	8257	8274	3 5 8	11 14
34	$\bar{1}$·8290	8306	8323	8339	8355	8371	8388	8404	8420	8436	3 5 8	11 14
35	$\bar{1}$·8452	8468	8484	8501	8517	8533	8549	8565	8581	8597	3 5 8	11 13
36	$\bar{1}$·8613	8629	8644	8660	8676	8692	8708	8724	8740	8755	3 5 8	11 13
37	$\bar{1}$·8771	8787	8803	8818	8834	8850	8865	8881	8897	8912	3 5 8	10 13
38	$\bar{1}$·8928	8944	8959	8975	8990	9006	9022	9037	9053	9068	3 5 8	10 13
39	$\bar{1}$·9084	9099	9115	9130	9146	9161	9176	9192	9207	9223	3 5 8	10 13
40°	$\bar{1}$·9238	9254	9269	9284	9300	9315	9330	9346	9361	9376	3 5 8	10 13
41	$\bar{1}$·9392	9407	9422	9438	9453	9468	9483	9499	9514	9529	3 5 8	10 13
42	$\bar{1}$·9544	9560	9575	9590	9605	9621	9636	9651	9666	9681	3 5 8	10 13
43	$\bar{1}$·9697	9712	9727	9742	9757	9773	9788	9803	9818	9833	3 5 8	10 13
44	$\bar{1}$·9848	9864	9879	9894	9909	9924	9939	9955	9970	9985	3 5 8	10 13

	0' 0·0°	6' 0·1°	12' 0·2°	18' 0·3°	24' 0·4°	30' 0·5°	36' 0·6°	42' 0·7°	48' 0·8°	54' 0·9°	Mean Differences				
											1'	2'	3'	4'	5'
45°	0·0000	0015	0030	0045	0061	0076	0091	0106	0121	0136	3	5	8	10	13
46	0·0152	0167	0182	0197	0212	0228	0243	0258	0273	0288	3	5	8	10	13
47	0·0303	0319	0334	0349	0364	0379	0395	0410	0425	0440	3	5	8	10	13
48	0·0456	0471	0486	0501	0517	0532	0547	0562	0578	0593	3	5	8	10	13
49	0·0608	0624	0639	0654	0670	0685	0700	0716	0731	0746	3	5	8	10	13
50°	0·0762	0777	0793	0808	0824	0839	0854	0870	0885	0901	3	5	8	10	13
51	0·0916	0932	0947	0963	0978	0994	1010	1025	1041	1056	3	5	8	10	13
52	0·1072	1088	1103	1119	1135	1150	1166	1182	1197	1213	3	5	8	10	13
53	0·1229	1245	1260	1276	1292	1308	1324	1340	1356	1371	3	5	8	11	13
54	0·1387	1403	1419	1435	1451	1467	1483	1499	1516	1532	3	5	8	11	13
55	0·1548	1564	1580	1596	1612	1629	1645	1661	1677	1694	3	5	8	11	14
56	0·1710	1726	1743	1759	1776	1792	1809	1825	1842	1858	3	5	8	11	14
57	0·1875	1891	1908	1925	1941	1958	1975	1992	2008	2025	3	6	8	11	14
58	0·2042	2059	2076	2093	2110	2127	2144	2161	2178	2195	3	6	9	11	14
59	0·2212	2229	2247	2264	2281	2299	2316	2333	2351	2368	3	6	9	12	14
60°	0·2386	2403	2421	2438	2456	2474	2491	2509	2527	2545	3	6	9	12	15
61	0·2562	2580	2598	2616	2634	2652	2670	2689	2707	2725	3	6	9	12	15
62	0·2743	2762	2780	2798	2817	2835	2854	2872	2891	2910	3	6	9	12	15
63	0·2928	2947	2966	2985	3004	3023	3042	3061	3080	3099	3	6	9	13	16
64	0·3118	3137	3157	3176	3196	3215	3235	3254	3274	3294	3	6	10	13	16
65	0·3313	3333	3353	3373	3393	3413	3433	3453	3473	3494	3	7	10	13	17
66	0·3514	3535	3555	3576	3596	3617	3638	3659	3679	3700	3	7	10	14	17
67	0·3721	3743	3764	3785	3806	3828	3849	3871	3892	3914	4	7	11	14	18
68	0·3936	3958	3980	4002	4024	4046	4068	4091	4113	4136	4	7	11	15	19
69	0·4158	4181	4204	4227	4250	4273	4296	4319	4342	4366	4	8	12	15	19
70°	0·4389	4413	4437	4461	4484	4509	4533	4557	4581	4606	4	8	12	16	20
71	0·4630	4655	4680	4705	4730	4755	4780	4805	4831	4857	4	8	13	17	21
72	0·4882	4908	4934	4960	4986	5013	5039	5066	5093	5120	4	9	13	18	22
73	0·5147	5174	5201	5229	5256	5284	5312	5340	5368	5397	5	9	14	19	23
74	0·5425	5454	5483	5512	5541	5570	5600	5629	5659	5689	5	10	15	20	25
75	0·5719	5750	5780	5811	5842	5873	5905	5936	5968	6000	5	10	16	21	26
76	0·6032	6065	6097	6130	6163	6196	6230	6264	6298	6332	6	11	17	22	28
77	0·6366	6401	6436	6471	6507	6542	6578	6615	6651	6688	6	12	18	24	30
78	0·6725	6763	6800	6838	6877	6915	6954	6994	7033	7073	6	13	19	26	32
79	0·7113	7154	7195	7236	7278	7320	7363	7406	7449	7493	7	14	21	28	35
80°	0·7537	7581	7626	7672	7718	7764	7811	7858	7906	7954	8	16	23	31	39
81	0·8003	8052	8102	8152	8203	8255	8307	8360	8413	8467	9	17	26	35	43
82	0·8522	8577	8633	8690	8748	8806	8865	8924	8985	9046	10	20	29	39	49
83	0·9109	9172	9236	9301	9367	9433	9501	9570	9640	9711	11	22	34	45	56
84	0·9784	9857	9932	$\overline{0}$008	$\overline{0}$085	$\overline{0}$164	$\overline{0}$244	$\overline{0}$326	$\overline{0}$409	$\overline{0}$494	13	26	40	53	66
85	1·0580	0669	0759	0850	0944	1040	1138	1238	1341	1446	16	32	48	64	81
86	1·1554	1664	1777	1893	2012	2135	2261	2391	2525	2663					
87	1·2806	2954	3106	3264	3429	3599	3777	3962	4155	4357					
88	1·4569	4792	5027	5275	5539	5819	6119	6441	6789	7167					
89	1·7581	8038	8550	9130	9800	$\overline{0}$591	$\overline{1}$561	$\overline{2}$810	$\overline{4}$571	7581					

	0'	6'	12'	18'	24'	30'	36'	42'	48'	54'
0°	0·0000	0017	0035	0052	0070	0087	0105	0122	0140	0157
1	0·0175	0192	0209	0227	0244	0262	0279	0297	0314	0332
2	0·0349	0367	0384	0401	0419	0436	0454	0471	0489	0506
3	0·0524	0541	0559	0576	0593	0611	0628	0646	0663	0681
4	0·0698	0716	0733	0750	0768	0785	0803	0820	0838	0855
5	0·0873	0890	0908	0925	0942	0960	0977	0995	1012	1030
6	0·1047	1065	1082	1100	1117	1134	1152	1169	1187	1204
7	0·1222	1239	1257	1274	1292	1309	1326	1344	1361	1379
8	0·1396	1414	1431	1449	1466	1484	1501	1518	1536	1553
9	0·1571	1588	1606	1623	1641	1658	1676	1693	1710	1728
10°	0·1745	1763	1780	1798	1815	1833	1850	1868	1885	1902
11	0·1920	1937	1955	1972	1990	2007	2025	2042	2059	2077
12	0·2094	2112	2129	2147	2164	2182	2199	2217	2234	2251
13	0·2269	2286	2304	2321	2339	2356	2374	2391	2409	2426
14	0·2443	2461	2478	2496	2513	2531	2548	2566	2583	2601
15	0·2618	2635	2653	2670	2688	2705	2723	2740	2758	2775
16	0·2793	2810	2827	2845	2862	2880	2897	2915	2932	2950
17	0·2967	2985	3002	3019	3037	3054	3072	3089	3107	3124
18	0·3142	3159	3176	3194	3211	3229	3246	3264	3281	3299
19	0·3316	3334	3351	3368	3386	3403	3421	3438	3456	3473
20°	0·3491	3508	3526	3543	3560	3578	3595	3613	3630	3648
21	0·3665	3683	3700	3718	3735	3752	3770	3787	3805	3822
22	0·3840	3857	3875	3892	3910	3927	3944	3962	3979	3997
23	0·4014	4032	4049	4067	4084	4102	4119	4136	4154	4171
24	0·4189	4206	4224	4241	4259	4276	4294	4311	4328	4346
25	0·4363	4381	4398	4416	4433	4451	4468	4485	4503	4520
26	0·4538	4555	4573	4590	4608	4625	4643	4660	4677	4695
27	0·4712	4730	4747	4765	4782	4800	4817	4835	4852	4869
28	0·4887	4904	4922	4939	4957	4974	4992	5009	5027	5044
29	0·5061	5079	5096	5114	5131	5149	5166	5184	5201	5219
30°	0·5236	5253	5271	5288	5306	5323	5341	5358	5376	5393
31	0·5411	5428	5445	5463	5480	5498	5515	5533	5550	5568
32	0·5585	5603	5620	5637	5655	5672	5690	5707	5725	5742
33	0·5760	5777	5794	5812	5829	5847	5864	5882	5899	5917
34	0·5934	5952	5969	5986	6004	6021	6039	6056	6074	6091
35	0·6109	6126	6144	6161	6178	6196	6213	6231	6248	6266
36	0·6283	6301	6318	6336	6353	6370	6388	6405	6423	6440
37	0·6458	6475	6493	6510	6528	6545	6562	6580	6597	6615
38	0·6632	6650	6667	6685	6702	6720	6737	6754	6772	6789
39	0·6807	6824	6842	6859	6877	6894	6912	6929	6946	6964
40°	0·6981	6999	7016	7034	7051	7069	7086	7103	7121	7138
41	0·7156	7173	7191	7208	7226	7243	7261	7278	7295	7313
42	0·7330	7348	7365	7383	7400	7418	7435	7453	7470	7487
43	0·7505	7522	7540	7557	7575	7592	7610	7627	7645	7662
44	0·7679	7697	7714	7732	7749	7767	7784	7802	7819	7837

Difference

for	is
1'	3
2'	6
3'	9
4'	12
5'	15

	0'	6'	12'	18'	24'	30'	36'	42'	48'	54'
45°	0·7854	7871	7889	7906	7924	7941	7959	7976	7994	8011
46	0·8029	8046	8063	8081	8098	8116	8133	8151	8168	8186
47	0·8203	8221	8238	8255	8273	8290	8308	8325	8343	8360
48	0·8378	8395	8412	8430	8447	8465	8482	8500	8517	8525
49	0·8552	8570	8587	8604	8622	8639	8657	8674	8692	8709
50°	0·8727	8744	8762	8779	8796	8814	8831	8849	8866	8884
51	0·8901	8919	8936	8954	8971	8988	9006	9023	9041	9058
52	0·9076	9093	9111	9128	9146	9163	9180	9198	9215	9233
53	0·9250	9268	9285	9303	9320	9338	9355	9372	9390	9407
54	0·9425	9442	9460	9477	9495	9512	9529	9547	9564	9582
55	0·9599	9617	9634	9652	9669	9687	9704	9721	9739	9756
56	0·9774	9791	9809	9826	9844	9861	9879	9896	9913	9931
57	0·9948	9966	9983	1·0001	1·0018	1·0036	1·0053	1·0071	1·0088	1·0105
58	1·0123	1·0140	1·0158	1·0175	1·0193	1·0210	1·0228	1·0245	1·0263	1·0280
59	1·0297	1·0315	1·0332	1·0350	1·0367	1·0385	1·0402	1·0420	1·0437	1·0455
60°	1·0472	1·0489	1·0507	1·0524	1·0542	1·0559	1·0577	1·0594	1·0612	1·0629
61	1·0647	1·0664	1·0681	1·0699	1·0716	1·0734	1·0751	1·0769	1·0786	1·0804
62	1·0821	1·0838	1·0856	1·0873	1·0891	1·0908	1·0926	1·0943	1·0961	1·0978
63	1·0996	1·1013	1·1030	1·1048	1·1065	1·1083	1·1100	1·1118	1·1135	1·1153
64	1·1170	1·1188	1·1205	1·1222	1·1240	1·1257	1·1275	1·1292	1·1310	1·1327
65	1·1345	1·1362	1·1380	1·1397	1·1414	1·1432	1·1449	1·1467	1·1484	1·1502
66	1·1519	1·1537	1·1554	1·1572	1·1589	1·1606	1·1624	1·1641	1·1659	1·1676
67	1·1694	1·1711	1·1729	1·1746	1·1764	1·1781	1·1798	1·1816	1·1833	1·1851
68	1·1868	1·1886	1·1903	1·1921	1·1938	1·1956	1·1973	1·1990	1·2008	1·2025
69	1·2043	1·2060	1·2078	1·2095	1·2113	1·2130	1·2147	1·2165	1·2182	1·2200
70°	1·2217	1·2235	1·2252	1·2270	1·2287	1·2305	1·2322	1·2339	1·2357	1 2374
71	1·2392	1·2409	1·2427	1·2444	1·2462	1·2479	1·2497	1·2514	1·2531	1·2549
72	1·2566	1·2584	1·2601	1·2619	1·2636	1·2654	1·2671	1·2689	1·2706	1·2723
73	1·2741	1·2758	1·2776	1·2793	1·2811	1·2828	1·2846	1·2863	1·2881	1·2898
74	1·2915	1·2933	1·2950	1·2968	1·2985	1·3003	1·3020	1·3038	1·3055	1·3073
75	1·3090	1·3107	1·3125	1·3142	1·3160	1·3177	1·3195	1·3212	1·3230	1·3247
76	1·3265	1·3282	1·3299	1·3317	1·3334	1·3352	1·3369	1·3387	1·3404	1·3422
77	1·3439	1·3456	1·3474	1·3491	1·3509	1·3526	1·3544	1·3561	1·3579	1·3596
78	1·3614	1·3631	1·3648	1·3666	1·3683	1·3701	1·3718	1·3736	1·3753	1·3771
79	1·3788	1·3806	1·3823	1·3840	1·3858	1·3875	1·3893	1·3910	1·3928	1·3945
80°	1·3963	1·3980	1·3998	1·4015	1·4032	1·4050	1·4067	1·4085	1·4102	1·4120
81	1·4137	1·4155	1·4172	1·4190	1·4207	1·4224	1·4242	1·4259	1·4277	1·4294
82	1·4312	1·4329	1·4347	1·4364	1·4382	1·4399	1·4416	1·4434	1·4451	1·4469
83	1·4486	1·4504	1·4521	1·4539	1·4556	1·4573	1·4591	1·4608	1·4626	1·4643
84	1·4661	1·4678	1·4696	1·4713	1·4731	1·4748	1·4765	1·4783	1·4800	1·4818
85	1·4835	1·4853	1·4870	1·4888	1·4905	1·4923	1·4940	1·4957	1·4975	1·4992
86	1·5010	1·5027	1·5045	1·5062	1·5080	1·5097	1·5115	1·5132	1·5149	1·5167
87	1·5184	1·5202	1·5219	1·5237	1·5254	1·5272	1·5289	1·5307	1·5324	1·5341
88	1·5359	1·5376	1·5394	1·5411	1·5429	1·5446	1·5464	1·5481	1·5499	1·5516
89	1·5533	1·5551	1·5568	1·5586	1·5603	1·5621	1·5638	1·5656	1·5673	1·5691

Difference

for	is
1'	3
2'	6
3'	9
4'	12
5'	15

Squares

	0	1	2	3	4	5	6	7	8	9
0	0	1	4	9	16	25	36	49	64	81
1	100	121	144	169	196	225	256	289	324	361
2	400	441	484	529	576	625	676	729	784	841
3	900	961	1024	1089	1156	1225	1296	1369	1444	1521
4	1600	1681	1764	1849	1936	2025	2116	2209	2304	2401
5	2500	2601	2704	2809	2916	3025	3136	3249	3364	3481
6	3600	3721	3844	3969	4096	4225	4356	4489	4624	4761
7	4900	5041	5184	5329	5476	5625	5776	5929	6084	6241
8	6400	6561	6724	6889	7056	7225	7396	7569	7744	7921
9	8100	8281	8464	8649	8836	9025	9216	9409	9604	9801
10	10000	10201	10404	10609	10816	11025	11236	11449	11664	11881
11	12100	12321	12544	12769	12996	13225	13456	13689	13924	14161
12	14400	14641	14884	15129	15376	15625	15876	16129	16384	16641
13	16900	17161	17424	17689	17956	18225	18496	18769	19044	19321
14	19600	19881	20164	20449	20736	21025	21316	21609	21904	22201
15	22500	22801	23104	23409	23716	24025	24336	24649	24964	25281
16	25600	25921	26244	26569	26896	27225	27556	27889	28224	28561
17	28900	29241	29584	29929	30276	30625	30976	31329	31684	32041
18	32400	32761	33124	33489	33856	34225	34596	34969	35344	35721
19	36100	36481	36864	37249	37636	38025	38416	38809	39204	39601
20	40000	40401	40804	41209	41616	42025	42436	42849	43264	43681
21	44100	44521	44944	45369	45796	46225	46656	47089	47524	47961
22	48400	48841	49284	49729	50176	50625	51076	51529	51984	52441
23	52900	53361	53824	54289	54756	55225	55696	56169	56644	57121
24	57600	58081	58564	59049	59536	60025	60516	61009	61504	62001
25	62500	63001	63504	64009	64516	65025	65536	66049	66564	67081
26	67600	68121	68644	69169	69696	70225	70756	71289	71824	72361
27	72900	73441	73984	74529	75076	75625	76176	76729	77284	77841
28	78400	78961	79524	80089	80656	81225	81796	82369	82944	83521
29	84100	84681	85264	85849	86436	87025	87616	88209	88804	89401
30	90000	90601	91204	91809	92416	93025	93636	94249	94864	95481
31	96100	96721	97344	97969	98596	99225	99856	100489	101124	101761
32	102400	103041	103684	104329	104976	105625	106276	106929	107584	108241
33	108900	109561	110224	110889	111556	112225	112896	113569	114244	114921
34	115600	116281	116964	117649	118336	119025	119716	120409	121104	121801
35	122500	123201	123904	124609	125316	126025	126736	127449	128164	128881
36	129600	130321	131044	131769	132496	133225	133956	134689	135424	136161
37	136900	137641	138384	139129	139876	140625	141376	142129	142884	143641
38	144400	145161	145924	146689	147456	148225	148996	149769	150544	151321
39	152100	152881	153664	154449	155236	156025	156816	157609	158404	159201
40	160000	160801	161604	162409	163216	164025	164836	165649	166464	167281
41	168100	168921	169744	170569	171396	172225	173056	173889	174724	175561
42	176400	177241	178084	178929	179776	180625	181476	182329	183184	184041
43	184900	185761	186624	187489	188356	189225	190096	190969	191844	192721
44	193600	194481	195364	196249	197136	198025	198916	199809	200704	201601
45	202500	203401	204304	205209	206116	207025	207936	208849	209764	210681
46	211600	212521	213444	214369	215296	216225	217156	218089	219024	219961
47	220900	221841	222784	223729	224676	225625	226576	227529	228484	229441
48	230400	231361	232324	233289	234256	235225	236196	237169	238144	239121
49	240100	241081	242064	243049	244036	245025	246016	247009	248004	249001

Exact squares of 4 figure numbers can be quickly calculated from the Identity
$$(a \pm b)^2 = a^2 \pm 2ab + b^2.$$

	0	1	2	3	4	5	6	7	8	9
50	250000	251001	252004	253009	254016	255025	256036	257049	258064	259081
51	260100	261121	262144	263169	264196	265225	266256	267289	268324	269361
52	270400	271441	272484	273529	274576	275625	276676	277729	278784	279841
53	280900	281961	283024	284089	285156	286225	287296	288369	289444	290521
54	291600	292681	293764	294849	295936	297025	298116	299209	300304	301401
55	302500	303601	304704	305809	306916	308025	309136	310249	311364	312481
56	313600	314721	315844	316969	318096	319225	320356	321489	322624	323761
57	324900	326041	327184	328329	329476	330625	331776	332929	334084	335241
58	336400	337561	338724	339889	341056	342225	343396	344569	345744	346921
59	348100	349281	350464	351649	352836	354025	355216	356409	357604	358801
60	360000	361201	362404	363609	364816	366025	367236	368449	369664	370881
61	372100	373321	374544	375769	376996	378225	379456	380689	381924	383161
62	384400	385641	386884	388129	389376	390625	391876	393129	394384	395641
63	396900	398161	399424	400689	401956	403225	404496	405769	407044	408321
64	409600	410881	412164	413449	414736	416025	417316	418609	419904	421201
65	422500	423801	425104	426409	427716	429025	430336	431649	432964	434281
66	435600	436921	438244	439569	440896	442225	443556	444889	446224	447561
67	448900	450241	451584	452929	454276	455625	456976	458329	459684	461041
68	462400	463761	465124	466489	467856	469225	470596	471969	473344	474721
69	476100	477481	478864	480249	481636	483025	484416	485809	487204	488601
70	490000	491401	492804	494209	495616	497025	498436	499849	501264	502681
71	504100	505521	506944	508369	509796	511225	512656	514089	515524	516961
72	518400	519841	521284	522729	524176	525625	527076	528529	529984	531441
73	532900	534361	535824	537289	538756	540225	541696	543169	544644	546121
74	547600	549081	550564	552049	553536	555025	556516	558009	559504	561001
75	562500	564001	565504	567009	568516	570025	571536	573049	574564	576081
76	577600	579121	580644	582169	583696	585225	586756	588289	589824	591361
77	592900	594441	595984	597529	599076	600625	602176	603729	605284	606841
78	608400	609961	611524	613089	614656	616225	617796	619369	620944	622521
79	624100	625681	627264	628849	630436	632025	633616	635209	636804	638401
80	640000	641601	643204	644809	646416	648025	649636	651249	652864	654481
81	656100	657721	659344	660969	662596	664225	665856	667489	669124	670761
82	672400	674041	675684	677329	678976	680625	682276	683929	685584	687241
83	688900	690561	692224	693889	695556	697225	698896	700569	702244	703921
84	705600	707281	708964	710649	712336	714025	715716	717409	719104	720801
85	722500	724201	725904	727609	729316	731025	732736	734449	736164	737881
86	739600	741321	743044	744769	746496	748225	749956	751689	753424	755161
87	756900	758641	760384	762129	763876	765625	767376	769129	770884	772641
88	774400	776161	777924	779689	781456	783225	784996	786769	788544	790321
89	792100	793881	795664	797449	799236	801025	802816	804609	806404	808201
90	810000	811801	813604	815409	817216	819025	820836	822649	824464	826281
91	828100	829941	831744	833569	835396	837225	839056	840889	842724	844561
92	846400	848241	850084	851929	853776	855625	857476	859329	861184	863041
93	864900	866761	868624	870489	872356	874225	876096	877969	879844	881721
94	883600	885481	887364	889249	891136	893025	894916	896809	898704	900601
95	902500	904401	906304	908209	910116	912025	913936	915849	917764	919681
96	921600	923521	925444	927369	929296	931225	933156	935089	937024	938961
97	940900	942841	944784	946729	948676	950625	952576	954529	956484	958441
98	960400	962361	964324	966289	968256	970225	972196	974169	976144	978121
99	980100	982081	984064	986049	988036	990025	992016	994009	996004	998001

Exact squares of 4 figure numbers can be quickly calculated from the identity
$$(a \pm b)^2 = a^2 \pm 2ab + b^2.$$

	0	1	2	3	4	5	6	7	8	9	Mean Differences								
											1	2	3	4	5	6	7	8	9
1·0	1·000	1·005	1·010	1·015	1·020	1·025	1·030	1·034	1·039	1·044	0	1	1	2	2	3	3	4	4
1·1	1·049	1·054	1·058	1·063	1·068	1·072	1·077	1·082	1·086	1·091	0	1	1	2	2	3	3	4	4
1·2	1·095	1·100	1·105	1·109	1·114	1·118	1·123	1·127	1·131	1·136	0	1	1	2	2	3	3	4	4
1·3	1·140	1·145	1·149	1·153	1·158	1·162	1·166	1·171	1·175	1·179	0	1	1	2	2	3	3	3	4
1·4	1·183	1·187	1·192	1·196	1·200	1·204	1·208	1·212	1·217	1·221	0	1	1	2	2	2	3	3	4
1·5	1·225	1·229	1·233	1·237	1·241	1·245	1·249	1·253	1·257	1·261	0	1	1	2	2	2	3	3	4
1·6	1·265	1·269	1·273	1·277	1·281	1·285	1·288	1·292	1·296	1·300	0	1	1	2	2	2	3	3	3
1·7	1·304	1·308	1·312	1·315	1·319	1·323	1·327	1·330	1·334	1·338	0	1	1	2	2	2	3	3	3
1·8	1·342	1·345	1·349	1·353	1·357	1·360	1·364	1·368	1·371	1·375	0	1	1	1	2	2	3	3	3
1·9	1·378	1·382	1·386	1·389	1·393	1·396	1·400	1·404	1·407	1·411	0	1	1	1	2	2	3	3	3
2·0	1·414	1·418	1·421	1·425	1·428	1·432	1·435	1·439	1·442	1·446	0	1	1	1	2	2	2	3	3
2·1	1·449	1·453	1·456	1·460	1·463	1·466	1·470	1·473	1·477	1·480	0	1	1	1	2	2	2	3	3
2·2	1·483	1·487	1·490	1·493	1·497	1·500	1·503	1·507	1·510	1·513	0	1	1	1	2	2	2	3	3
2·3	1·517	1·520	1·523	1·526	1·530	1·533	1·536	1·539	1·543	1·546	0	1	1	1	2	2	2	3	3
2·4	1·549	1·552	1·556	1·559	1·562	1·565	1·568	1·572	1·575	1·578	0	1	1	1	2	2	2	3	3
2·5	1·581	1·584	1·587	1·591	1·594	1·597	1·600	1·603	1·606	1·609	0	1	1	1	2	2	2	3	3
2·6	1·612	1·616	1·619	1·622	1·625	1·628	1·631	1·634	1·637	1·640	0	1	1	1	2	2	2	2	3
2·7	1·643	1·646	1·649	1·652	1·655	1·658	1·661	1·664	1·667	1·670	0	1	1	1	2	2	2	2	3
2·8	1·673	1·676	1·679	1·682	1·685	1·688	1·691	1·694	1·697	1·700	0	1	1	1	1	2	2	2	3
2·9	1·703	1·706	1·709	1·712	1·715	1·718	1·720	1·723	1·726	1·729	0	1	1	1	1	2	2	2	3
3·0	1·732	1·735	1·738	1·741	1·744	1·746	1·749	1·752	1·755	1·758	0	1	1	1	1	2	2	2	3
3·1	1·761	1·764	1·766	1·769	1·772	1·775	1·778	1·780	1·783	1·786	0	1	1	1	1	2	2	2	3
3·2	1·789	1·792	1·794	1·797	1·800	1·803	1·806	1·808	1·811	1·814	0	1	1	1	1	2	2	2	2
3·3	1·817	1·819	1·822	1·825	1·828	1·830	1·833	1·836	1·839	1·841	0	1	1	1	1	2	2	2	2
3·4	1·844	1·847	1·849	1·852	1·855	1·857	1·860	1·863	1·866	1·868	0	1	1	1	1	2	2	2	2
3·5	1·871	1·874	1·876	1·879	1·882	1·884	1·887	1·889	1·892	1·895	0	1	1	1	1	2	2	2	2
3·6	1·897	1·900	1·903	1·905	1·908	1·911	1·913	1·916	1·918	1·921	0	1	1	1	1	2	2	2	2
3·7	1·924	1·926	1·929	1·931	1·934	1·937	1·939	1·942	1·944	1·947	0	1	1	1	1	2	2	2	2
3·8	1·949	1·952	1·955	1·957	1·960	1·962	1·965	1·967	1·970	1·972	0	1	1	1	1	2	2	2	2
3·9	1·975	1·977	1·980	1·982	1·985	1·988	1·990	1·993	1·995	1·998	0	1	1	1	1	2	2	2	2
4·0	2·000	2·003	2·005	2·008	2·010	2·013	2·015	2·017	2·020	2·022	0	0	1	1	1	1	2	2	2
4·1	2·025	2·027	2·030	2·032	2·035	2·037	2·040	2·042	2·045	2·047	0	0	1	1	1	1	2	2	2
4·2	2·049	2·052	2·054	2·057	2·059	2·062	2·064	2·066	2·069	2·071	0	0	1	1	1	1	2	2	2
4·3	2·074	2·076	2·078	2·081	2·083	2·086	2·088	2·091	2·093	2·095	0	0	1	1	1	1	2	2	2
4·4	2·098	2·100	2·102	2·105	2·107	2·110	2·112	2·114	2·117	2·119	0	0	1	1	1	1	2	2	2
4·5	2·121	2·124	2·126	2·128	2·131	2·133	2·135	2·138	2·140	2·142	0	0	1	1	1	1	2	2	2
4·6	2·145	2·147	2·149	2·152	2·154	2·156	2·159	2·161	2·163	2·166	0	0	1	1	1	1	2	2	2
4·7	2·168	2·170	2·173	2·175	2·177	2·179	2·182	2·184	2·186	2·189	0	0	1	1	1	1	2	2	2
4·8	2·191	2·193	2·195	2·198	2·200	2·202	2·205	2·207	2·209	2·211	0	0	1	1	1	1	2	2	2
4·9	2·214	2·216	2·218	2·220	2·223	2·225	2·227	2·229	2·232	2·234	0	0	1	1	1	1	2	2	2
5·0	2·236	2·238	2·241	2·243	2·245	2·247	2·249	2·252	2·254	2·256	0	0	1	1	1	1	2	2	2
5·1	2·258	2·261	2·263	2·265	2·267	2·269	2·272	2·274	2·276	2·278	0	0	1	1	1	1	2	2	2
5·2	2·280	2·283	2·285	2·287	2·289	2·291	2·294	2·296	2·298	2·300	0	0	1	1	1	1	2	2	2
5·3	2·302	2·304	2·307	2·309	2·311	2·313	2·315	2·317	2·320	2·322	0	0	1	1	1	1	2	2	2
5·4	2·324	2·326	2·328	2·330	2·332	2·335	2·337	2·339	2·341	2·343	0	0	1	1	1	1	1	2	2

	0	1	2	3	4	5	6	7	8	9	Mean Differences 1 2 3 4 5 6 7 8 9
5·5	2·345	2·347	2·350	2·352	2·354	2·356	2·358	2·360	2·362	2·364	0 0 1 1 1 1 1 2 2
5·6	2·366	2·369	2·371	2·373	2·375	2·377	2·379	2·381	2·383	2·385	0 0 1 1 1 1 1 2 2
5·7	2·388	2·390	2·392	2·394	2·396	2·398	2·400	2·402	2·404	2·406	0 0 1 1 1 1 1 2 2
5·8	2·408	2·410	2·412	2·415	2·417	2·419	2·421	2·423	2·425	2·427	0 0 1 1 1 1 1 2 2
5·9	2·429	2·431	2·433	2·435	2·437	2·439	2·441	2·443	2·445	2·447	0 0 1 1 1 1 1 2 2
6·0	2·450	2·452	2·454	2·456	2·458	2·460	2·462	2·464	2·466	2·468	0 0 1 1 1 1 1 2 2
6·1	2·470	2·472	2·474	2·476	2·478	2·480	2·482	2·484	2·486	2·488	0 0 1 1 1 1 1 2 2
6·2	2·490	2·492	2·494	2·496	2·498	2·500	2·502	2·504	2·506	2·508	0 0 1 1 1 1 1 2 2
6·3	2·510	2·512	2·514	2·516	2·518	2·520	2·522	2·524	2·526	2·528	0 0 1 1 1 1 1 2 2
6·4	2·530	2·532	2·534	2·536	2·538	2·540	2·542	2·544	2·546	2·548	0 0 1 1 1 1 1 2 2
6·5	2·550	2·551	2·553	2·555	2·557	2·559	2·561	2·563	2·565	2·567	0 0 1 1 1 1 1 2 2
6·6	2·569	2·571	2·573	2·575	2·577	2·579	2·581	2·583	2·585	2·587	0 0 1 1 1 1 1 2 2
6·7	2·588	2·590	2·592	2·594	2·596	2·598	2·600	2·602	2·604	2·606	0 0 1 1 1 1 1 2 2
6·8	2·608	2·610	2·612	2·613	2·615	2·617	2·619	2·621	2·623	2·625	0 0 1 1 1 1 1 2 2
6·9	2·627	2·629	2·631	2·632	2·634	2·636	2·638	2·640	2·642	2·644	0 0 1 1 1 1 1 2 2
7·0	2·646	2·648	2·650	2·651	2·653	2·655	2·657	2·659	2·661	2·663	0 0 1 1 1 1 1 2 2
7·1	2·665	2·667	2·668	2·670	2·672	2·674	2·676	2·678	2·680	2·681	0 0 1 1 1 1 1 1 2
7·2	2·683	2·685	2·687	2·689	2·691	2·693	2·694	2·696	2·698	2·700	0 0 1 1 1 1 1 1 2
7·3	2·702	2·704	2·706	2·707	2·709	2·711	2·713	2·715	2·717	2·719	0 0 1 1 1 1 1 1 2
7·4	2·720	2·722	2·724	2·726	2·728	2·729	2·731	2·733	2·735	2·737	0 0 1 1 1 1 1 1 2
7·5	2·739	2·740	2·742	2·744	2·746	2·748	2·750	2·751	2·753	2·755	0 0 1 1 1 1 1 1 2
7·6	2·757	2·759	2·760	2·762	2·764	2·766	2·768	2·769	2·771	2·773	0 0 1 1 1 1 1 1 2
7·7	2·775	2·777	2·778	2·780	2·782	2·784	2·786	2·787	2·789	2·791	0 0 1 1 1 1 1 1 2
7·8	2·793	2·795	2·796	2·798	2·800	2·802	2·804	2·805	2·807	2·809	0 0 1 1 1 1 1 1 2
7·9	2·811	2·812	2·814	2·816	2·818	2·820	2·821	2·823	2·825	2·827	0 0 1 1 1 1 1 1 2
8·0	2·828	2·830	2·832	2·834	2·835	2·837	2·839	2·841	2·843	2·844	0 0 1 1 1 1 1 1 2
8·1	2·846	2·848	2·850	2·851	2·853	2·855	2·857	2·858	2·860	2·862	0 0 1 1 1 1 1 1 2
8·2	2·864	2·865	2·867	2·869	2·871	2·872	2·874	2·876	2·877	2·879	0 0 1 1 1 1 1 1 2
8·3	2·881	2·883	2·884	2·886	2·888	2·890	2·891	2·893	2·895	2·897	0 0 1 1 1 1 1 1 2
8·4	2·898	2·900	2·902	2·903	2·905	2·907	2·909	2·910	2·912	2·914	0 0 1 1 1 1 1 1 2
8·5	2·915	2·917	2·919	2·921	2·922	2·924	2·926	2·927	2·929	2·931	0 0 1 1 1 1 1 1 2
8·6	2·933	2·934	2·936	2·938	2·939	2·941	2·943	2·944	2·946	2·948	0 0 1 1 1 1 1 1 2
8·7	2·950	2·951	2·953	2·955	2·956	2·958	2·960	2·961	2·963	2·965	0 0 1 1 1 1 1 1 2
8·8	2·966	2·968	2·970	2·972	2·973	2·975	2·977	2·978	2·980	2·982	0 0 1 1 1 1 1 1 2
8·9	2·983	2·985	2·987	2·988	2·990	2·992	2·993	2·995	2·997	2·998	0 0 1 1 1 1 1 1 2
9·0	3·000	3·002	3·003	3·005	3·007	3·008	3·010	3·012	3·013	3·015	0 0 0 1 1 1 1 1 1
9·1	3·017	3·018	3·020	3·022	3·023	3·025	3·027	3·028	3·030	3·032	0 0 0 1 1 1 1 1 1
9·2	3·033	3·035	3·036	3·038	3·040	3·041	3·043	3·045	3·046	3·048	0 0 0 1 1 1 1 1 1
9·3	3·050	3·051	3·053	3·055	3·056	3·058	3·059	3·061	3·063	3·064	0 0 0 1 1 1 1 1 1
9·4	3·066	3·068	3·069	3·071	3·072	3·074	3·076	3·077	3·079	3·081	0 0 0 1 1 1 1 1 1
9·5	3·082	3·084	3·085	3·087	3·089	3·090	3·092	3·094	3·095	3·097	0 0 0 1 1 1 1 1 1
9·6	3·098	3·100	3·102	3·103	3·105	3·106	3·108	3·110	3·111	3·113	0 0 0 1 1 1 1 1 1
9·7	3·115	3·116	3·118	3·119	3·121	3·123	3·124	3·126	3·127	3·129	0 0 0 1 1 1 1 1 1
9·8	3·131	3·132	3·134	3·135	3·137	3·139	3·140	3·142	3·143	2·145	0 0 0 1 1 1 1 1 1
9·9	3·146	3·148	3·150	3·151	3·153	3·154	3·156	3·158	3·159	3·161	0 0 0 1 1 1 1 1 1

	0	1	2	3	4	5	6	7	8	9	Mean Differences								
											1	2	3	4	5	6	7	8	9
10	3·162	3·178	3·194	3·209	3·225	3·240	3·256	3·271	3·286	3·302	2	3	5	6	8	9	11	12	14
11	3·317	3·332	3·347	3·362	3·376	3·391	3·406	3·421	3·435	3·450	1	3	4	6	7	9	10	12	13
12	3·464	3·479	3·493	3·507	3·521	3·536	3·550	3·564	3·578	3·592	1	3	4	6	7	8	10	11	13
13	3·606	3·619	3·633	3·647	3·661	3·674	3·688	3·701	3·715	3·728	1	3	4	5	7	8	10	11	12
14	3·742	3·755	3·768	3·782	3·795	3·808	3·821	3·834	3·847	3·860	1	3	4	5	7	8	9	11	12
15	3·873	3·886	3·899	3·912	3·924	3·937	3·950	3·962	3·975	3·988	1	3	4	5	6	8		9	10	11
16	4·000	4·012	4·025	4·037	4·050	4·062	4·074	4·087	4·099	4·111	1	2	4	5	6	7	9	10	11
17	4·123	4·135	4·147	4·159	4·171	4·183	4·195	4·207	4·219	4·231	1	2	4	5	6	7	8	10	11
18	4·243	4·254	4·266	4·278	4·290	4·301	4·313	4·324	4·336	4·347	1	2	3	5	6	7	8	9	10
19	4·359	4·370	4·382	4·393	4·405	4·416	4·427	4·438	4·450	4·461	1	2	3	5	6	7	8	9	10
20	4·472	4·483	4·494	4·506	4·517	4·528	4·539	4·550	4·561	4·572	1	2	3	4	6	7	8	9	10
21	4·583	4·594	4·604	4·615	4·626	4·637	4·648	4·658	4·669	4·680	1	2	3	4	5	6	8	9	10
22	4·690	4·701	4·712	4·722	4·733	4·743	4·754	4·765	4·775	4·785	1	2	3	4	5	6	7	8	9
23	4·796	4·806	4·817	4·827	4·837	4·848	4·858	4·868	4·879	4·889	1	2	3	4	5	6	7	8	9
24	4·899	4·909	4·919	4·930	4·940	4·950	4·960	4·970	4·980	4·990	1	2	3	4	5	6	7	8	9
25	5·000	5·010	5·020	5·030	5·040	5·050	5·060	5·070	5·079	5·089	1	2	3	4	5	6	7	8	9
26	5·099	5·109	5·119	5·128	5·138	5·148	5·158	5·167	5·177	5·187	1	2	3	4	5	6	7	8	9
27	5·196	5·206	5·215	5·225	5·235	5·244	5·254	5·263	5·273	5·282	1	2	3	4	5	6	7	8	9
28	5·292	5·301	5·310	5·320	5·329	5·339	5·348	5·357	5·367	5·376	1	2	3	4	5	6	7	7	8
29	5·385	5·394	5·404	5·413	5·422	5·431	5·441	5·450	5·459	5·468	1	2	3	4	5	5	6	7	8
30	5·477	5·486	5·495	5·505	5·514	5·523	5·532	5·541	5·550	5·559	1	2	3	4	4	5	6	7	8
31	5·568	5·577	5·586	5·595	5·604	5·612	5·621	5·630	5·639	5·648	1	2	3	3	4	5	6	7	8
32	5·657	5·666	5·675	5·683	5·692	5·701	5·710	5·718	5·727	5·736	1	2	3	3	4	5	6	7	8
33	5·745	5·753	5·762	5·771	5·779	5·788	5·797	5·805	5·814	5·822	1	2	3	3	4	5	6	7	8
34	5·831	5·840	5·848	5·857	5·865	5·874	5·882	5·891	5·899	5·908	1	2	3	3	4	5	6	7	8
35	5·916	5·925	5·933	5·941	5·950	5·958	5·967	5·975	5·983	5·992	1	2	2	3	4	5	6	7	8
36	6·000	6·008	6·017	6·025	6·033	6·042	6·050	6·058	6·066	6·075	1	2	2	3	4	5	6	7	7
37	6·083	6·091	6·099	6·107	6·116	6·124	6·132	6·140	6·148	6·156	1	2	2	3	4	5	6	7	7
38	6·164	6·173	6·181	6·189	6·197	6·205	6·213	6·221	6·229	6·237	1	2	2	3	4	5	6	6	7
39	6·245	6·253	6·261	6·269	6·277	6·285	6·293	6·301	6·309	6·317	1	2	2	3	4	5	6	6	7
40	6·325	6·332	6·340	6·348	6·356	6·364	6·372	6·380	6·387	6·395	1	2	2	3	4	5	6	6	7
41	6·403	6·411	6·419	6·427	6·434	6·442	6·450	6·458	6·465	6·473	1	2	2	3	4	5	5	6	7
42	6·481	6·488	6·496	6·504	6·512	6·519	6·527	6·535	6·542	6·550	1	2	2	3	4	5	5	6	7
43	6·557	6·565	6·573	6·580	6·588	6·595	6·603	6·611	6·618	6·626	1	2	2	3	4	5	5	6	7
44	6·633	6·641	6·648	6·656	6·663	6·671	6·678	6·686	6·693	6·701	1	2	2	3	4	5	5	6	7
45	6·708	6·716	6·723	6·731	6·738	6·745	6·753	6·760	6·768	6·775	1	1	2	3	4	4	5	6	7
46	6·782	6·790	6·797	6·804	6·812	6·819	6·826	6·834	6·841	6·848	1	1	2	3	4	4	5	6	7
47	6·856	6·863	6·870	6·878	6·885	6·892	6·899	6·907	6·914	6·921	1	1	2	3	4	4	5	6	7
48	6·928	6·935	6·943	6·950	6·957	6·964	6·971	6·979	6·986	6·993	1	1	2	3	4	4	5	6	6
49	7·000	7·007	7·014	7·021	7·029	7·036	7·043	7·050	7·057	7·064	1	1	2	3	4	4	5	6	6
50	7·071	7·078	7·085	7·092	7·099	7·106	7·113	7·120	7·127	7·134	1	1	2	3	4	4	5	6	6
51	7·141	7·148	7·155	7·162	7·169	7·176	7·183	7·190	7·197	7·204	1	1	2	3	4	4	5	6	6
52	7·211	7·218	7·225	7·232	7·239	7·246	7·253	7·259	7·266	7·273	1	1	2	3	3	4	5	6	6
53	7·280	7·287	7·294	7·301	7·308	7·314	7·321	7·328	7·335	7·342	1	1	2	3	3	4	5	5	6
54	7·349	7·355	7·362	7·369	7·376	7·382	7·389	7·396	7·403	7·410	1	1	2	3	3	4	5	5	6

	0	1	2	3	4	5	6	7	8	9	Mean Differences 1 2 3 4 5 6 7 8 9
55	7·416	7·423	7·430	7·436	7·443	7·450	7·457	7·463	7·470	7·477	1 1 2 3 3 4 5 5 6
56	7·483	7·490	7·497	7·503	7·510	7·517	7·523	7·530	7·537	7·543	1 1 2 3 3 4 5 5 6
57	7·550	7·556	7·563	7·570	7·576	7·583	7·589	7·596	7·603	7·609	1 1 2 3 3 4 5 5 6
58	7·616	7·622	7·629	7·635	7·642	7·649	7·655	7·662	7·668	7·675	1 1 2 3 3 4 5 5 6
59	7·681	7·688	7·694	7·701	7·707	7·714	7·720	7·727	7·733	7·740	1 1 2 3 3 4 4 5 6
60	7·746	7·752	7·759	7·765	7·772	7·778	7·785	7·791	7·797	7·804	1 1 2 3 3 4 4 5 6
61	7·810	7·817	7·823	7·829	7·836	7·842	7·849	7·855	7·861	7·868	1 1 2 3 3 4 4 5 6
62	7·874	7·880	7·887	7·893	7·899	7·906	7·912	7·918	7·925	7·931	1 1 2 3 3 4 4 5 6
63	7·937	7·944	7·950	7·956	7·962	7·969	7·975	7·981	7·987	7·994	1 1 2 3 3 4 4 5 6
64	8·000	8·006	8·012	8·019	8·025	8·031	8·037	8·044	8·050	8·056	1 1 2 2 3 4 4 5 6
65	8·062	8·068	8·075	8·081	8·087	8·093	8·099	8·106	8·112	8·118	1 1 2 2 3 4 4 5 6
66	8·124	8·130	8·136	8·142	8·149	8·155	8·161	8·167	8·173	8·179	1 1 2 2 3 4 4 5 5
67	8·185	8·191	8·198	8·204	8·210	8·216	8·222	8·228	8·234	8·240	1 1 2 2 3 4 4 5 5
68	8·246	8·252	8·258	8·264	8·270	8·276	8·283	8·289	8·295	8·301	1 1 2 2 3 4 4 5 5
69	8·307	8·313	8·319	8·325	8·331	8·337	8·343	8·349	8·355	8·361	1 1 2 2 3 4 4 5 5
70	8·367	8·373	8·379	8·385	8·390	8·396	8·402	8·408	8·414	8·420	1 1 2 2 3 4 4 5 5
71	8·426	8·432	8·438	8·444	8·450	8·456	8·462	8·468	8·473	8·479	1 1 2 2 3 4 4 5 5
72	8·485	8·491	8·497	8·503	8·509	8·515	8·521	8·526	8·532	8·538	1 1 2 2 3 3 4 5 5
73	8·544	8·550	8·556	8·562	8·567	8·573	8·579	8·585	8·591	8·597	1 1 2 2 3 3 4 5 5
74	8·602	8·608	8·614	8·620	8·626	8·631	8·637	8·643	8·649	8·654	1 1 2 2 3 3 4 5 5
75	8·660	8·666	8·672	8·678	8·683	8·689	8·695	8·701	8·706	8·712	1 1 2 2 3 3 4 5 5
76	8·718	8·724	8·729	8·735	8·741	8·746	8·752	8·758	8·764	8·769	1 1 2 2 3 3 4 4 5
77	8·775	8·781	8·786	8·792	8·798	8·803	8·809	8·815	8·820	8·826	1 1 2 2 3 3 4 4 5
78	8·832	8·837	8·843	8·849	8·854	8·860	8·866	8·871	8·877	8·883	1 1 2 2 3 3 4 4 5
79	8·888	8·894	8·899	8·905	8·911	8·916	8·922	8·927	8·933	8·939	1 1 2 2 3 3 4 4 5
80	8·944	8·950	8·955	8·961	8·967	8·972	8·978	8·983	8·989	8·994	1 1 2 2 3 3 4 4 5
81	9·000	9·006	9·011	9·017	9·022	9·028	9·033	9·039	9·044	9·050	1 1 2 2 3 3 4 4 5
82	9·055	9·061	9·066	9·072	9·077	9·083	9·088	9·094	9·099	9·105	1 1 2 2 3 3 4 4 5
83	9·110	9·116	9·121	9·127	9·132	9·138	9·143	9·149	9·154	9·160	1 1 2 2 3 3 4 4 5
84	9·165	9·171	9·176	9·182	9·187	9·192	9·198	9·203	9·209	9·214	1 1 2 2 3 3 4 4 5
85	9·220	9·225	9·230	9·236	9·241	9·247	9·252	9·257	9·263	9·268	1 1 2 2 3 3 4 4 5
86	9·274	9·279	9·284	9·290	9·295	9·301	9·306	9·311	9·317	9·322	1 1 2 2 3 3 4 4 5
87	9·327	9·333	9·338	9·343	9·349	9·354	9·359	9·365	9·370	9·375	1 1 2 2 3 3 4 4 5
88	9·381	9·386	9·391	9·397	9·402	9·407	9·413	9·418	9·423	9·429	1 1 2 2 3 3 4 4 5
89	9·434	9·439	9·445	9·450	9·455	9·460	9·466	9·471	9·476	9·482	1 1 2 2 3 3 4 4 5
90	9·487	9·492	9·497	9·503	9·508	9·513	9·518	9·524	9·529	9·534	1 1 2 2 3 3 4 4 5
91	9·539	9·545	9·550	9·555	9·560	9·566	9·571	9·576	9·581	9·586	1 1 2 2 3 3 4 4 5
92	9·592	9·597	9·602	9·607	9·613	9·618	9·623	9·628	9·633	9·638	1 1 2 2 3 3 4 4 5
93	9·644	9·649	9·654	9·659	9·664	9·670	9·675	9·680	9·685	9·690	1 1 2 2 3 3 4 4 5
94	9·695	9·701	9·706	9·711	9·716	9·721	9·726	9·731	9·737	9·742	1 1 2 2 3 3 4 4 5
95	9·747	9·752	9·757	9·762	9·767	9·772	9·778	9·783	9·788	9·793	1 1 2 2 3 3 4 4 5
96	9·798	9·803	9·808	9·813	9·818	9·823	9·829	9·834	9·839	9·844	1 1 2 2 3 3 4 4 5
97	9·849	9·854	9·859	9·864	9·869	9·874	9·879	9·884	9·889	9·894	1 1 1 2 3 3 4 4 5
98	9·900	9·905	9·910	9·915	9·920	9·925	9·930	9·935	9·940	9·945	0 1 1 2 2 3 3 4 4
99	9·950	9·955	9·960	9·965	9·970	9·975	9·980	9·985	9·990	9·995	0 1 1 2 2 3 3 4 4

	0	1	2	3	4	5	6	7	8	9	SUBTRACT Mean Differences								
											1	2	3	4	5	6	7	8	9
1·0	1·000	9901	9804	9709	9615	9524	9434	9346	9259	9174	9	18	28	37	46	55	64	74	83
1·1	0·9091	9009	8929	8850	8772	8696	8621	8547	8475	8403	8	15	23	31	38	46	53	61	69
1·2	0·8333	8264	8197	8130	8065	8000	7937	7874	7813	7752	7	13	20	26	33	39	46	52	59
1·3	0·7692	7634	7576	7519	7463	7407	7353	7299	7246	7194	6	11	17	22	28	33	39	44	50
1·4	0·7143	7092	7042	6993	6944	6897	6849	6803	6757	6711	5	10	14	19	24	29	33	38	43
1·5	0·6667	6623	6579	6536	6494	6452	6410	6369	6329	6289	4	8	13	17	21	25	29	33	38
1·6	0·6250	6211	6173	6135	6098	6061	6024	5988	5952	5917	4	7	11	15	18	22	26	29	33
1·7	0·5882	5848	5814	5780	5747	5714	5682	5650	5618	5587	3	7	10	13	16	20	23	26	29
1·8	0·5556	5525	5495	5464	5435	5405	5376	5348	5319	5291	3	6	9	12	15	17	20	23	26
1·9	0·5263	5236	5208	5181	5155	5128	5102	5076	5051	5025	3	5	8	11	13	16	18	21	24
2·0	0·5000	4975	4950	4926	4902	4878	4854	4831	4808	4785	2	5	7	10	12	14	17	19	21
2·1	0·4762	4739	4717	4695	4673	4651	4630	4608	4587	4566	2	4	7	9	11	13	15	17	19
2·2	0·4545	4525	4505	4484	4464	4444	4425	4405	4386	4367	2	4	6	8	10	12	14	16	18
2·3	0·4348	4329	4310	4292	4274	4255	4237	4219	4202	4184	2	4	5	7	9	11	13	14	16
2·4	0·4167	4149	4132	4115	4098	4082	4065	4049	4032	4016	2	3	5	7	8	10	12	13	15
2·5	0·4000	3984	3968	3953	3937	3922	3906	3891	3876	3861	2	3	5	6	8	9	11	12	14
2·6	0·3846	3831	3817	3802	3788	3774	3759	3745	3731	3717	1	3	4	6	7	9	10	11	13
2·7	0·3704	3690	3676	3663	3650	3636	3623	3610	3597	3584	1	3	4	5	7	8	9	11	12
2·8	0·3571	3559	3546	3534	3521	3509	3497	3484	3472	3460	1	2	4	5	6	7	9	10	11
2·9	0·3448	3436	3425	3413	3401	3390	3378	3367	3356	3344	1	2	3	5	6	7	8	9	10
3·0	0·3333	3322	3311	3300	3289	3279	3268	3257	3247	3236	1	2	3	4	5	6	8	9	10
3·1	0·3226	3215	3205	3195	3185	3175	3165	3155	3145	3135	1	2	3	4	5	6	7	8	9
3·2	0·3125	3115	3106	3096	3086	3077	3067	3058	3049	3040	1	2	3	4	5	6	7	8	9
3·3	0·3030	3021	3012	3003	2994	2985	2976	2967	2959	2950	1	2	3	4	4	5	6	7	8
3·4	0·2941	2933	2924	2915	2907	2899	2890	2882	2874	2865	1	2	3	3	4	5	6	7	8
3·5	0·2857	2849	2841	2833	2825	2817	2809	2801	2793	2786	1	2	2	3	4	5	6	6	7
3·6	0·2778	2770	2762	2755	2747	2740	2732	2725	2717	2710	1	2	2	3	4	5	5	6	7
3·7	0·2703	2695	2688	2681	2674	2667	2660	2653	2646	2639	1	1	2	3	4	4	5	6	6
3·8	0·2632	2625	2618	2611	2604	2597	2591	2584	2577	2571	1	1	2	3	3	4	5	5	6
3·9	0·2564	2558	2551	2545	2538	2532	2525	2519	2513	2506	1	1	2	3	3	4	4	5	6
4·0	0·2500	2494	2488	2481	2475	2469	2463	2457	2451	2445	1	1	2	2	3	4	4	5	5
4·1	0·2439	2433	2427	2421	2415	2410	2404	2398	2392	2387	1	1	2	2	3	3	4	5	5
4·2	0·2381	2375	2370	2364	2358	2353	2347	2342	2336	2331	1	1	2	2	3	3	4	4	5
4·3	0·2326	2320	2315	2309	2304	2299	2294	2288	2283	2278	1	1	2	2	3	3	4	4	5
4·4	0·2273	2268	2262	2257	2252	2247	2242	2237	2232	2227	1	1	2	2	3	3	4	4	5
4·5	0·2222	2217	2212	2208	2203	2198	2193	2188	2183	2179	0	1	1	2	2	3	3	4	4
4·6	0·2174	2169	2165	2160	2155	2151	2146	2141	2137	2132	0	1	1	2	2	3	3	4	4
4·7	0·2128	2123	2119	2114	2110	2105	2101	2096	2092	2088	0	1	1	2	2	3	3	4	4
4·8	0·2083	2079	2075	2070	2066	2062	2058	2053	2049	2045	0	1	1	2	2	3	3	3	4
4·9	0·2041	2037	2033	2028	2024	2020	2016	2012	2008	2004	0	1	1	2	2	2	3	3	4
5·0	0·2000	1996	1992	1988	1984	1980	1976	1972	1969	1965	0	1	1	2	2	2	3	3	4
5·1	0·1961	1957	1953	1949	1946	1942	1938	1934	1931	1927	0	1	1	2	2	2	3	3	3
5·2	0·1923	1919	1916	1912	1908	1905	1901	1898	1894	1890	0	1	1	1	2	2	3	3	3
5·3	0·1887	1883	1880	1876	1873	1869	1866	1862	1859	1855	0	1	1	1	2	2	2	3	3
5·4	0·1852	1848	1845	1842	1838	1835	1832	1828	1825	1821	0	1	1	1	2	2	2	3	3

e.g. $\dfrac{1}{3\cdot7} = 0\cdot2703$, $\dfrac{1}{3\cdot74} = 0\cdot2674$, $\dfrac{1}{3\cdot748} = 0\cdot2668$, $\dfrac{1}{374\cdot8} = 0\cdot002668$, $\dfrac{1}{0\cdot0003748} = 2668$.

	0	1	2	3	4	5	6	7	8	9	SUBTRACT Mean Differences								
											1	2	3	4	5	6	7	8	9
5·5	0·1818	1815	1812	1808	1805	1802	1799	1795	1792	1789	0	1	1	1	2	2	2	3	3
5·6	0·1786	1783	1779	1776	1773	1770	1767	1764	1761	1757	0	1	1	1	2	2	2	3	3
5·7	0·1754	1751	1748	1745	1742	1739	1736	1733	1730	1727	0	1	1	1	2	2	2	2	3
5·8	0·1724	1721	1718	1715	1712	1709	1706	1704	1701	1698	0	1	1	1	1	2	2	2	3
5·9	0·1695	1692	1689	1686	1684	1681	1678	1675	1672	1669	0	1	1	1	1	2	2	2	3
6·0	0·1667	1664	1661	1658	1656	1653	1650	1647	1645	1642	0	1	1	1	1	2	2	2	3
6·1	0·1639	1637	1634	1631	1629	1626	1623	1621	1618	1616	0	1	1	1	1	2	2	2	2
6·2	0·1613	1610	1608	1605	1603	1600	1597	1595	1592	1590	0	1	1	1	1	2	2	2	2
6·3	0·1587	1585	1582	1580	1577	1575	1572	1570	1567	1565	0	0	1	1	1	1	2	2	2
6·4	0·1562	1560	1558	1555	1553	1550	1548	1546	1543	1541	0	0	1	1	1	1	2	2	2
6·5	0·1538	1536	1534	1531	1529	1527	1524	1522	1520	1517	0	0	1	1	1	1	2	2	2
6·6	0·1515	1513	1511	1508	1506	1504	1502	1499	1497	1495	0	0	1	1	1	1	2	2	2
6·7	0·1493	1490	1488	1486	1484	1481	1479	1477	1475	1473	0	0	1	1	1	1	2	2	2
6·8	0·1471	1468	1466	1464	1462	1460	1458	1456	1453	1451	0	0	1	1	1	1	2	2	2
6·9	0·1449	1447	1445	1443	1441	1439	1437	1435	1433	1431	0	0	1	1	1	1	2	2	2
7·0	0·1429	1427	1425	1422	1420	1418	1416	1414	1412	1410	0	0	1	1	1	1	1	2	2
7·1	0·1408	1406	1404	1403	1401	1399	1397	1395	1393	1391	0	0	1	1	1	1	1	2	2
7·2	0·1389	1387	1385	1383	1381	1379	1377	1376	1374	1372	0	0	1	1	1	1	1	2	2
7·3	0·1370	1368	1366	1364	1362	1361	1359	1357	1355	1353	0	0	1	1	1	1	1	2	2
7·4	0·1351	1350	1348	1346	1344	1342	1340	1339	1337	1335	0	0	1	1	1	1	1	1	2
7·5	0·1333	1332	1330	1328	1326	1325	1323	1321	1319	1318	0	0	1	1	1	1	1	1	2
7·6	0·1316	1314	1312	1311	1309	1307	1305	1304	1302	1300	0	0	1	1	1	1	1	1	2
7·7	0·1299	1297	1295	1294	1292	1290	1289	1287	1285	1284	0	0	0	1	1	1	1	1	1
7·8	0·1282	1280	1279	1277	1276	1274	1272	1271	1269	1267	0	0	0	1	1	1	1	1	1
7·9	0·1266	1264	1263	1261	1259	1258	1256	1255	1253	1252	0	0	0	1	1	1	1	1	1
8·0	0·1250	1248	1247	1245	1244	1242	1241	1239	1238	1236	0	0	0	1	1	1	1	1	1
8·1	0·1235	1233	1232	1230	1229	1227	1225	1224	1222	1221	0	0	0	1	1	1	1	1	1
8·2	0·1220	1218	1217	1215	1214	1212	1211	1209	1208	1206	0	0	0	1	1	1	1	1	1
8·3	0·1205	1203	1202	1200	1199	1198	1196	1195	1193	1192	0	0	0	1	1	1	1	1	1
8·4	0·1190	1189	1188	1186	1185	1183	1182	1181	1179	1178	0	0	0	1	1	1	1	1	1
8·5	0·1176	1175	1174	1172	1171	1170	1168	1167	1166	1164	0	0	0	1	1	1	1	1	1
8·6	0·1163	1161	1160	1159	1157	1156	1155	1153	1152	1151	0	0	0	1	1	1	1	1	1
8·7	0·1149	1148	1147	1145	1144	1143	1142	1140	1139	1138	0	0	0	1	1	1	1	1	1
8·8	0·1136	1135	1134	1133	1131	1130	1129	1127	1126	1125	0	0	0	1	1	1	1	1	1
8·9	0·1124	1122	1121	1120	1119	1117	1116	1115	1114	1112	0	0	0	1	1	1	1	1	1
9·0	0·1111	1110	1109	1107	1106	1105	1104	1103	1101	1100	0	0	0	1	1	1	1	1	1
9·1	0·1099	1098	1096	1095	1094	1093	1092	1090	1089	1088	0	0	0	0	1	1	1	1	1
9·2	0·1087	1086	1085	1083	1082	1081	1080	1079	1078	1076	0	0	0	0	1	1	1	1	1
9·3	0·1075	1074	1073	1072	1071	1070	1068	1067	1066	1065	0	0	0	0	1	1	1	1	1
9·4	0·1064	1063	1062	1060	1059	1058	1057	1056	1055	1054	0	0	0	0	1	1	1	1	1
9·5	0·1053	1052	1050	1049	1048	1047	1046	1045	1044	1043	0	0	0	0	1	1	1	1	1
9·6	0·1042	1041	1039	1038	1037	1036	1035	1034	1033	1032	0	0	0	0	1	1	1	1	1
9·7	0·1031	1030	1029	1028	1027	1026	1025	1024	1022	1021	0	0	0	0	1	1	1	1	1
9·8	0·1020	1019	1018	1017	1016	1015	1014	1013	1012	1011	0	0	0	0	1	1	1	1	1
9·9	0·1010	1009	1008	1007	1006	1005	1004	1003	1002	1001	0	0	0	0	0	1	1	1	1

Natural Logarithms

	0	1	2	3	4	5	6	7	8	9	1	2	3	4	5	6	7	8	9
1·0	0·0000	0100	0198	0296	0392	0488	0583	0677	0770	0862	10	19	29	38	48	57	67	76	86
1·1	0·0953	1044	1133	1222	1310	1398	1484	1570	1655	1740	9	17	26	35	44	52	61	70	78
1·2	0·1823	1906	1989	2070	2151	2231	2311	2390	2469	2546	8	16	24	32	40	48	56	64	72
1·3	0·2624	2700	2776	2852	2927	3001	3075	3148	3221	3293	7	15	22	30	37	44	52	59	67
1·4	0·3365	3436	3507	3577	3646	3716	3784	3853	3920	3988	7	14	21	28	35	41	48	55	62
1·5	0·4055	4121	4187	4253	4318	4383	4447	4511	4574	4637	6	13	19	26	32	39	45	52	58
1·6	0·4700	4762	4824	4886	4947	5008	5068	5128	5188	5247	6	12	18	24	30	36	42	48	55
1·7	0·5306	5365	5423	5481	5539	5596	5653	5710	5766	5822	6	11	17	23	29	34	40	46	51
1·8	0·5878	5933	5988	6043	6098	6152	6206	6259	6313	6366	5	11	16	22	27	32	38	43	49
1·9	0·6419	6471	6523	6575	6627	6678	6729	6780	6831	6881	5	10	15	20	26	31	36	41	46
2·0	0·6931	6981	7031	7080	7129	7178	7227	7275	7324	7372	5	10	15	20	24	29	34	39	44
2·1	0·7419	7467	7514	7561	7608	7655	7701	7747	7793	7839	5	9	14	19	23	28	33	37	42
2·2	0·7885	7930	7975	8020	8065	8109	8154	8198	8242	8286	4	9	13	18	22	27	31	36	40
2·3	0·8329	8372	8416	8459	8502	8544	8587	8629	8671	8713	4	9	13	17	21	26	30	34	38
2·4	0·8755	8796	8838	8879	8920	8961	9002	9042	9083	9123	4	8	12	16	20	24	29	33	37
2·5	0·9163	9203	9243	9282	9322	9361	9400	9439	9478	9517	4	8	12	16	20	24	27	31	35
2·6	0·9555	9594	9632	9670	9708	9746	9783	9821	9858	9895	4	8	11	15	19	23	26	30	34
2·7	0·9933	9969	0̄006	0̄043	0̄080	0̄116	0̄152	0̄188	0̄225	0̄260	4	7	11	15	18	22	25	29	33
2·8	1·0296	0332	0367	0403	0438	0473	0508	0543	0578	0613	4	7	11	14	18	21	25	28	32
2·9	1·0647	0682	0716	0750	0784	0818	0852	0886	0919	0953	3	7	10	14	17	20	24	27	31
3·0	1·0986	1019	1053	1086	1119	1151	1184	1217	1249	1282	3	7	10	13	16	20	23	26	30
3·1	1·1314	1346	1378	1410	1442	1474	1506	1537	1569	1600	3	6	10	13	16	19	22	25	29
3·2	1·1632	1663	1694	1725	1756	1787	1817	1848	1878	1909	3	6	9	12	15	18	22	25	28
3·3	1·1939	1969	2000	2030	2060	2090	2119	2149	2179	2208	3	6	9	12	15	18	21	24	27
3·4	1·2238	2267	2296	2326	2355	2384	2413	2442	2470	2499	3	6	9	12	15	17	20	23	26
3·5	1·2528	2556	2585	2613	2641	2669	2698	2726	2754	2782	3	6	8	11	14	17	20	23	25
3·6	1·2809	2837	2865	2892	2920	2947	2975	3002	3029	3056	3	5	8	11	14	16	19	22	25
3·7	1·3083	3110	3137	3164	3191	3218	3244	3271	3297	3324	3	5	8	11	13	16	19	21	24
3·8	1·3350	3376	3403	3429	3455	3481	3507	3533	3558	3584	3	5	8	10	13	16	18	21	23
3·9	1·3610	3635	3661	3686	3712	3737	3762	3788	3813	3838	3	5	8	10	13	15	18	20	23
4·0	1·3863	3888	3913	3938	3962	3987	4012	4036	4061	4085	2	5	7	10	12	15	17	20	22
4·1	1·4110	4134	4159	4183	4207	4231	4255	4279	4303	4327	2	5	7	10	12	14	17	19	22
4·2	1·4351	4375	4398	4422	4446	4469	4493	4516	4540	4563	2	5	7	9	12	14	16	19	21
4·3	1·4586	4609	4633	4656	4679	4702	4725	4748	4770	4793	2	5	7	9	12	14	16	18	21
4·4	1·4816	4839	4861	4884	4907	4929	4953	4974	4996	5019	2	5	7	9	11	14	16	18	20
4·5	1·5041	5063	5085	5107	5129	5151	5173	5195	5217	5239	2	4	7	9	11	13	15	18	20
4·6	1·5261	5282	5304	5326	5347	5369	5390	5412	5433	5454	2	4	6	9	11	13	15	17	19
4·7	1·5476	5497	5518	5539	5560	5581	5602	5623	5644	5665	2	4	6	8	11	13	15	17	19
4·8	1·5686	5707	5728	5748	5769	5790	5810	5831	5851	5872	2	4	6	8	10	12	14	16	19
4·9	1·5892	5913	5933	5953	5974	5994	6014	6034	6054	6074	2	4	6	8	10	12	14	16	18
5·0	1·6094	6114	6134	6154	6174	6194	6214	6233	6253	6273	2	4	6	8	10	12	14	16	18
5·1	1·6292	6312	6332	6351	6371	6390	6409	6429	6448	6467	2	4	6	8	10	12	14	16	18
5·2	1·6487	6506	6525	6544	6563	6582	6601	6620	6639	6658	2	4	6	8	10	11	13	15	17
5·3	1·6677	6696	6715	6734	6752	6771	6790	6808	6827	6845	2	4	6	7	9	11	13	15	17
5·4	1·6864	6882	6901	6919	6938	6956	6974	6993	7011	7029	2	4	5	7	9	11	13	15	17

NATURAL LOGARITHMS OF 10^n

n	1	2	3	4	5	6	7	8	9
$\log_e 10^n$	2·3026	4·6052	6·9078	9·2103	11·5129	13·8155	16·1181	18·4207	20·7233

E.g. $\log_e 584·7 = \log_e(5·847 \times 10^2) = 1·7659 + 4·6052 = 6·3711$.

	0	1	2	3	4	5	6	7	8	9	1	2	3	4	5	6	7	8	9
5·5	1·7047	7066	7084	7102	7120	7138	7156	7174	7192	7210	2	4	5	7	9	11	13	14	16
5·6	1·7228	7246	7263	7281	7299	7317	7334	7352	7370	7387	2	4	5	7	9	11	12	14	16
5·7	1·7405	7422	7440	7457	7475	7492	7509	7527	7544	7561	2	3	5	7	9	10	12	14	16
5·8	1·7579	7596	7613	7630	7647	7664	7681	7699	7716	7733	2	3	5	7	9	10	12	14	15
5·9	1·7750	7766	7783	7800	7817	7834	7851	7867	7884	7901	2	3	5	7	8	10	12	13	15
6·0	1·7918	7934	7951	7967	7984	8001	8017	8034	8050	8066	2	3	5	7	8	10	12	13	15
6·1	1·8083	8099	8116	8132	8148	8165	8181	8197	8213	8229	2	3	5	6	8	10	11	13	15
6·2	1·8245	8262	8278	8294	8310	8326	8342	8358	8374	8390	2	3	5	6	8	10	11	13	14
6·3	1·8405	8421	8437	8453	8469	8485	8500	8516	8532	8547	2	3	5	6	8	9	11	13	14
6·4	1·8563	8579	8594	8610	8625	8641	8656	8672	8687	8703	2	3	5	6	8	9	11	12	14
6·5	1·8718	8733	8749	8764	8779	8795	8810	8825	8840	8856	2	3	5	6	8	9	11	12	14
6·6	1·8871	8886	8901	8916	8931	8946	8961	8976	8991	9006	2	3	5	6	8	9	11	12	14
6·7	1·9021	9036	9051	9066	9081	9095	9110	9125	9140	9155	1	3	4	6	7	9	10	12	13
6·8	1·9169	9184	9199	9213	9228	9242	9257	9272	9286	9301	1	3	4	6	7	9	10	12	13
6·9	1·9315	9330	9344	9359	9373	9387	9402	9416	9430	9445	1	3	4	6	7	9	10	12	13
7·0	1·9459	9473	9488	9502	9516	9530	9544	9559	9573	9587	1	3	4	6	7	9	10	11	13
7·1	1·9601	9615	9629	9643	9657	9671	9685	9699	9713	9727	1	3	4	6	7	8	10	11	13
7·2	1·9741	9755	9769	9782	9796	9810	9824	9838	9851	9865	1	3	4	6	7	8	10	11	12
7·3	1·9879	9892	9906	9920	9933	9947	9961	9974	9988	0001	1	3	4	5	7	8	10	11	12
7·4	2·0015	0028	0042	0055	0069	0082	0096	0109	0122	0136	1	3	4	5	7	8	9	11	12
7·5	2·0149	0162	0176	0189	0202	0215	0229	0242	0255	0268	1	3	4	5	7	8	9	11	12
7·6	2·0282	0295	0308	0321	0334	0347	0360	0373	0386	0399	1	3	4	5	7	8	9	10	12
7·7	2·0412	0425	0438	0451	0464	0477	0490	0503	0516	0528	1	3	4	5	6	8	9	10	12
7·8	2·0541	0554	0567	0580	0592	0605	0618	0631	0643	0656	1	3	4	5	6	8	9	10	11
7·9	2·0669	0681	0694	0707	0719	0732	0744	0757	0769	0782	1	3	4	5	6	8	9	10	11
8·0	2·0794	0807	0819	0832	0844	0857	0869	0882	0894	0906	1	3	4	5	6	7	9	10	11
8·1	2·0919	0931	0943	0956	0968	0980	0992	1005	1017	1029	1	2	4	5	6	7	9	10	11
8·2	2·1041	1054	1066	1078	1090	1102	1114	1126	1138	1150	1	2	4	5	6	7	9	10	11
8·3	2·1163	1175	1187	1199	1211	1223	1235	1247	1258	1270	1	2	4	5	6	7	8	10	11
8·4	2·1282	1294	1306	1318	1330	1342	1353	1365	1377	1389	1	2	4	5	6	7	8	10	11
8·5	2·1401	1412	1424	1436	1448	1459	1471	1483	1494	1506	1	2	4	5	6	7	8	9	11
8·6	2·1518	1529	1541	1552	1564	1576	1587	1599	1610	1622	1	2	3	5	6	7	8	9	10
8·7	2·1633	1645	1656	1668	1679	1691	1702	1713	1725	1736	1	2	3	5	6	7	8	9	10
8·8	2·1748	1759	1770	1782	1793	1804	1815	1827	1838	1849	1	2	3	5	6	7	8	9	10
8·9	2·1861	1872	1883	1894	1905	1917	1928	1939	1950	1961	1	2	3	4	6	7	8	9	10
9·0	2·1972	1983	1994	2006	2017	2028	2039	2050	2061	2072	1	2	3	4	6	7	8	9	10
9·1	2·2083	2094	2105	2116	2127	2138	2148	2159	2170	2181	1	2	3	4	5	7	8	9	10
9·2	2·2192	2203	2214	2225	2235	2246	2257	2268	2279	2289	1	2	3	4	5	6	8	9	10
9·3	2·2300	2311	2322	2332	2343	2354	2364	2375	2386	2396	1	2	3	4	5	6	7	9	10
9·4	2·2407	2418	2428	2439	2450	2460	2471	2481	2492	2502	1	2	3	4	5	6	7	8	10
9·5	2·2513	2523	2534	2544	2555	2565	2576	2586	2597	2607	1	2	3	4	5	6	7	8	9
9·6	2·2618	2628	2638	2649	2659	2670	2680	2690	2701	2711	1	2	3	4	5	6	7	8	9
9·7	2·2721	2732	2742	2752	2762	2773	2783	2793	2803	2814	1	2	3	4	5	6	7	8	9
9·8	2·2824	2834	2844	2854	2865	2875	2885	2895	2905	2915	1	2	3	4	5	6	7	8	9
9·9	2·2925	2935	2946	2956	2966	2976	2986	2996	3006	3016	1	2	3	4	5	6	7	8	9

NATURAL LOGARITHMS OF 10^{-n}

n	1	2	3	4	5	6	7	8	9
$\log_e 10^{-n}$	$\bar{3}·6974$	$\bar{5}·3948$	$\bar{7}·0922$	$\overline{10}·7897$	$\overline{12}·4871$	$\overline{14}·1845$	$\overline{17}·8819$	$\overline{19}·5793$	$\overline{21}·2767$

E.g. $\log_e 0·005847 = \log_e(5·847 \times 10^{-3}) = 1·7659 + \bar{7}·0922 = \bar{6}·8581$

Exponential and Hyperbolic Functions

x	e^x	e^{-x}	$\theta°$ (gd x)	cosh x (sec θ)	sinh x (tan θ)	tanh x (sin θ)	log cosh x	log sinh x
0·1	1·1052	0·9048	5·720	1·0050	0·1002	0·0997	0·0022	$\bar{1}$·0007
0·2	1·2214	0·8187	11·384	1·0201	0·2013	0·1974	0·0086	$\bar{1}$·3039
0·3	1·3499	0·7408	16·937	1·0453	0·3045	0·2913	0·0193	$\bar{1}$·8436
0·4	1·4918	0·6703	22·331	1·0811	0·4108	0·3799	0·0339	$\bar{1}$·6136
0·5	1·6487	0·6065	27·524	1·1276	0·5211	0·4621	0·0522	$\bar{1}$·7169
0·6	1·8221	0·5488	32·483	1·1855	0·6367	0·5370	0·0739	$\bar{1}$·8093
0·7	2·0138	0·4966	37·183	1·2552	0·7586	0·6044	0·0987	$\bar{1}$·8800
0·8	2·2255	0·4493	41·608	1·3374	0·8881	0·6640	0·1263	$\bar{1}$·9485
0·9	2·4596	0·4066	45·750	1·4331	1·0265	0·7163	0·1563	0·0114
1·0	2·7183	0·3679	49·605	1·5431	1·1752	0·7616	0·1884	0·0701
1·1	3·0042	0·3329	53·178	1·6685	1·3356	0·8005	0·2223	0·1257
1·2	3·3201	0·3012	56·476	1·8107	1·5095	0·8337	0·2578	0·1788
1·3	3·6693	0·2725	59·511	1·9709	1·6984	0·8617	0·2947	0·2300
1·4	4·0552	0·2466	62·295	2·1509	1·9043	0·8854	0·3326	0·2797
1·5	4·4817	0·2231	64·843	2·3524	2·1293	0·9051	0·3715	0·3282
1·6	4·9530	0·2019	67·171	2·5775	2·3756	0·9217	0·4112	0·3758
1·7	5·4739	0·1827	69·294	2·8283	2·6456	0·9354	0·4515	0·4225
1·8	6·0496	0·1653	71·228	3·1075	2·9422	0·9468	0·4924	0·4687
1·9	6·6859	0·1496	72·987	3·4177	3·2682	0·9562	0·5337	0·5143
2·0	7·3891	0·1353	74·584	3·7622	3·6269	0·9640	0·5754	0·5595
2·1	8·1662	0·1225	76·037	4·1443	4·0219	0·9705	0·6175	0·6044
2·2	9·0250	0·1108	77·354	4·5679	4·4571	0·9757	0·6597	0·6491
2·3	9·9742	0·1003	78·549	5·0372	4·9370	0·9801	0·7022	0·6935
2·4	11·023	0·0907	79·633	5·5569	5·4662	0·9837	0·7448	0·7377
2·5	12·183	0·0821	80·615	6·1323	6·0502	0·9866	0·7876	0·7818
2·6	13·464	0·0743	81·513	6·7690	6·6947	0·9890	0·8305	0·8257
2·7	14·880	0·0672	82·310	7·4735	7·4063	0·9910	0·8735	0·8696
2·8	16·445	0·0608	83·040	8·2527	8·1919	0·9926	0·9166	0·9134
2·9	18·174	0·0550	83·707	9·1146	9·0596	0·9940	0·9597	0·9571
3·0	20·086	0·0498	84·301	10·068	10·018	0·9951	1·0029	1·0008
3·1	22·198	0·0450	84·841	11·121	11·076	0·9959	1·0462	1·0444
3·2	24·533	0·0408	85·336	12·287	12·246	0·9967	1·0894	1·0880
3·3	27·113	0·0369	85·775	13·575	13·538	0·9973	1·1327	1·1316
3·4	29·964	0·0334	86·177	14·999	14·965	0·9978	1·1761	1·1751
3·5	33·115	0·0302	86·541	16·573	16·543	0·9982	1·2194	1·2186
3·6	36·598	0·0273	86·870	18·313	18·285	0·9985	1·2628	1·2621
3·7	40·447	0·0247	87·168	20·236	20·211	0·9988	1·3061	1·3056
3·8	44·701	0·0224	87·437	22·362	22·339	0·9990	1·3495	1·3491
3·9	49·402	0·0202	87·681	24·711	24·691	0·9992	1·3929	1·3925
4·0	54·598	0·0183	87·903	27·308	27·290	0·9993	1·4363	1·4360
4·1	60·340	0·0166	88·104	30·178	30·162	0·9995	1·4797	1·4795
4·2	66·686	0·0150	88·281	33·351	33·336	0·9996	1·5231	1·5229
4·3	73·700	0·0136	88·447	36·857	36·843	0·9996	1·5665	1·5664
4·4	81·451	0·0123	88·591	40·732	40·719	0·9997	1·6099	1·6098
4·5	90·017	0·0111	88·728	45·014	45·003	0·9997	1·6533	1·6532
4·6	99·484	0·0101	88·849	49·747	49·737	0·9998	1·6968	1·6967
4·7	109·95	0·0091	88·957	54·978	54·969	0·9998	1·7402	1·7401
4·8	121·51	0·0082	89·055	60·759	60·751	0·9999	1·7836	1·7836
4·9	134·29	0·0074	89·146	67·149	67·141	0·9999	1·8270	1·8270
5·0	148·41	0·0067	89·227	74·210	74·203	0·9999	1·8705	1·8704

Mathematical Constants and Formulae

Constants

e = Base of natural logarithms $\approx 2\cdot71828$

$\log_{10}e \approx 0\cdot434294$ $\qquad \log_e10 \approx 2\cdot30259$

$\log_{10}N \approx \log_eN\times0\cdot4343$ $\qquad \log_e N \approx \log_{10}N\times2\cdot3026$

1 radian $\approx 57°\cdot2958 \approx 57° \ 17' \ 45''$ $\qquad \pi \approx 3\cdot14159265$

$\log_{10}\pi \approx 0\cdot49715$ $\quad \dfrac{1}{\pi} \approx 0\cdot31831$ $\quad \dfrac{\pi}{180} \approx 0\cdot01745$ $\quad \pi^2 \approx 9\cdot8696$

Algebra

$\log_a x = y \Leftrightarrow x = a^y$ $\quad \log_q p = \log_q r \log_r p$

Sum of first n terms of the series $a, a+d, a+2d, \ldots$

$S_n = \frac{1}{2}n[2a+(n-1)d] = n\times$(average of first and last terms)

$$\sum_{r=1}^{n} r = \tfrac{1}{2}n(n+1) \qquad \sum_{r=1}^{n} r^2 = \tfrac{1}{6}n(n+1)(2n+1)$$

$$\sum_{-1}^{n} r^3 = \tfrac{1}{4}n^2(n+1)^2 \qquad \sum_{r=0}^{n-1} a^r = \frac{1-a^n}{1-a}$$

If $f(x) \equiv ax^2+bx+c$, roots α, β of $f(x) = 0$

given by $\dfrac{(-b\pm\sqrt{b^2-4ac})}{2a}$ $\qquad$ Also $\alpha+\beta = \dfrac{-b}{a}$, $\alpha\beta = \dfrac{c}{a}$

$f(x) > 0$ all real $x <=> a > 0$, $c > 0$, $4ac > b^2$

Remainder when polynominal $P(x)$ divided by $(x-a)$ is $P(a)$

Number of combinations of n objects taken r at a time

$_nC$, or $\dbinom{n}{r} = \dfrac{n!}{(n-r)!r!}$ where $n! = n(n-1)(n-2) \ldots 3. \ 2. \ 1$

Complex number $Z = x+iy = r(\cos\theta+i\sin\theta) = re^{i\theta}$

Mod $Z = |Z| = r = \sqrt{x^2+y^2}$ $\qquad$ Arg $Z = \theta+2p\pi$

Where p is taken such that $-\pi < \text{Arg } Z \leqslant \pi$

$Z_1Z_2 = r_1 e^{i\theta_1} r_2 e^{i\theta_2} = r_1 r_2 e^{i(\theta_1+\theta_2)}$ $\qquad Z^n = r^n e^{in\theta}$

Vectors

If $\mathbf{x}$ has components $(x_1\,x_2, x_3)$ and $\mathbf{y}$ has components

(y_1, y_2, y_3) $\mathbf{x}.\mathbf{y} = x_1\,y_1+x_2\,y_2+x_3\,y_3$

and $\mathbf{x}\times\mathbf{y}$ has components $(x_2y_3-x_3y_2, \ x_3y_1-x_1y_3, \ x_1y_2-x_2y_1)$

$\mathbf{x}\times\mathbf{y}.\mathbf{z} = \begin{vmatrix} x_1 & x_2 & x_3 \\ y_1 & y_2 & y_3 \\ z_1 & z_2 & z_3 \end{vmatrix}$ $\qquad \mathbf{x}\times(\mathbf{y}\times\mathbf{z}) = (\mathbf{x}.\mathbf{z})\mathbf{y}-(\mathbf{x}.\mathbf{y})\mathbf{z}$

$\qquad\qquad\qquad\qquad\quad \nabla = \mathbf{i}\dfrac{\partial}{\partial x}+\mathbf{j}\dfrac{\partial}{\partial y}+\mathbf{k}\dfrac{\partial}{\partial z}$

Grad$\varphi = \nabla\varphi$ $\qquad$ Div $\mathbf{a} = \nabla.\mathbf{a}$ $\qquad$ Curl $\mathbf{a} = \nabla\times\mathbf{a}$

$$\nabla^2\varphi = \frac{\partial^2\varphi}{\partial x^2}+\frac{\partial^2\varphi}{\partial y^2}+\frac{\partial^2\varphi}{\partial z^2}$$

and in spherical polar coordinates r, θ, ψ

$$\nabla^2\varphi = \frac{\partial^2\varphi}{\partial r^2}+\frac{2}{r}\frac{\partial\varphi}{\partial r}+\frac{1}{r^2}\frac{\partial^2\varphi}{\partial\theta^2}+\frac{\cot\theta}{r^2}\frac{\partial\varphi}{\partial\theta}+\frac{1}{r^2\sin^2\theta}\frac{\partial^2\varphi}{\partial\psi^2}$$

Mathematical Constants and Formulae (Cont.)

Expansions and Approximations

Taylor's expansion: $f(a+x) = f(a) + xf'(a) + \dfrac{x^2}{2!}f''(a) + \dfrac{x^3}{3!}f'''(a) + \ldots$

or (Maclaurin's form): $f(x) = f(0) + xf'(0) + \dfrac{x^2}{2!}f''(0) + \dfrac{x^3}{3!}f'''(0) + \ldots$

Expansions (*valid if $|x| < 1$, the rest valid for all x)

$\sin x = \dfrac{x}{1!} - \dfrac{x^3}{3!} + \dfrac{x^5}{5!} - \dfrac{x^7}{7!} + \ldots$

$\cos x = 1 - \dfrac{x^2}{2!} + \dfrac{x^4}{4!} - \dfrac{x^6}{6!} + \ldots$

$e^x = 1 + \dfrac{x}{1!} + \dfrac{x^2}{2!} + \dfrac{x^3}{3!} + \ldots$

$\sinh x = x + \dfrac{x^3}{3!} + \dfrac{x^5}{5!} + \dfrac{x^7}{7!} + \ldots$

$\cosh x = 1 + \dfrac{x^2}{2!} + \dfrac{x^4}{4!} + \dfrac{x^6}{6!} + \ldots$

$*\log(1+x) = x - \dfrac{x^2}{2} + \dfrac{x^3}{3} - \dfrac{x^4}{4} + \ldots$

$*(1+x)^n = 1 + nx + \dfrac{n(n-1)}{2!}x^2 + \dfrac{n(n-1)(n-2)}{3!}x^3 + \ldots + \binom{n}{r}x^r + \ldots$

Newton–Raphson iterative formula for root of $f(x) = 0$: $x_{n+1} = x_n - \dfrac{f(x_n)}{f'(x_n)}$

Kinematics and Centres of Gravity

Components of acceleration in polar coordinates

Radial ... $a_r = \ddot{r} - r\dot{\theta}^2$ Tangential ... $a_t = \dfrac{1}{r}\dfrac{d}{dt}(r^2\dot{\theta})$

Distance of centre of gravity from the centre (d)

(a) Hemisphere (radius r) $d = \dfrac{3r}{8}$

(b) Hemispherical shell $d = \dfrac{r}{2}$

(c) Sector of circle (angle 2θ) $d = \dfrac{(2r\sin\theta)}{3\theta}$

(d) Arc of circle $d = \dfrac{(r\sin\theta)}{\theta}$

(e) Cone (height h) $d = \dfrac{h}{4}$ (from centre of base)

Analysis

(a) List of derivatives

y	$\dfrac{\mathrm{d}y}{\mathrm{d}x}$	y	$\dfrac{\mathrm{d}y}{\mathrm{d}x}$
$\sin x$	$\cos x$	$\cos x$	$-\sin x$
$\tan x$	$\sec^2 x$	$\cot x$	$-\operatorname{cosec}^2 x$
$\sec x$	$\sec x \tan x$	$\operatorname{cosec} x$	$-\operatorname{cosec} x \cot x$

(b) List of integrals

$F'(x)=f(x)$	$F(x)=\int f(x)\mathrm{d}x$	$F'(x)=f(x)$	$F(x)=\int f(x)\mathrm{d}x$
x^a	$\dfrac{x^{a+1}}{a+1}\quad a\neq -1$	$\dfrac{1}{a^2+x^2}$	$\dfrac{1}{a}\tan^{-1}\dfrac{x}{a}$
$\dfrac{1}{x}$	$\log|x|$	$\dfrac{1}{x\sqrt{x^2-a^2}}$	$\dfrac{1}{a}\sec^{-1}\dfrac{x}{a}$
e^x	e^x	$\dfrac{1}{(a^2-x^2)}$	$\dfrac{1}{a}\tanh^{-1}\dfrac{x}{a}$
a^x	$\dfrac{a^x}{\log a}$		$=\dfrac{1}{2a}\log\!\left(\dfrac{a+x}{a-x}\right)$
$\dfrac{1}{\sqrt{a^2-x^2}}$	$\sin^{-1}\dfrac{x}{a}$	$\tan x$	$\log|\sec x|$
$\dfrac{1}{\sqrt{x^2+a^2}}$	$\sinh^{-1}\dfrac{x}{a}$	$\sec x$	$\log|\sec x+\tan x|$
	$=\log\!\left(\dfrac{x}{a}+\sqrt{\dfrac{x^2}{a^2}+1}\right)$	$\operatorname{cosec} x$	$=\log\left|\tan\!\left(\dfrac{x}{2}+\dfrac{\pi}{4}\right)\right|$ $\log\left|\tan\dfrac{x}{2}\right|$
$\pm\dfrac{1}{\sqrt{x^2-a^2}}$	$\cosh^{-1}\dfrac{x}{a}$	$e^{ax}\sin(bx+c)$	$\dfrac{e^{ax}}{a^2+b^2}\Big[a\sin(bx+c)$
	$=\log\!\left(\dfrac{x}{a}\pm\sqrt{\dfrac{x^2}{a^2}-1}\right)$		$\qquad -b\cos(bx+c)\Big]$

Simpson's rule $\displaystyle\int_a^b f(x)\mathrm{d}x \approx \tfrac{1}{3}h(y_0+4y_1+y_2)$ where $h=\tfrac{1}{2}(b-a)$

$(uv)' = u'v+uv',\quad \left(\dfrac{u}{v}\right)' = \dfrac{u'v-uv'}{v^2}$

$\int uv'\,\mathrm{d}x = uv-\int u'v\,dx$

$\displaystyle\int_0^{\frac{\pi}{2}}\sin^m x\,\cos^n x\,\mathrm{d}x = \dfrac{(n-1)(n-3)\ldots(m-1)(m-3)}{(m+n)(m+n-2)}\ldots\lambda$

where $\lambda = \dfrac{\pi}{2}$ if $m,\,n$ both even, and 1 otherwise

Radius of curvature $\rho = \dfrac{\mathrm{d}s}{\mathrm{d}\psi} = \dfrac{\left[1+\left(\dfrac{\mathrm{d}y}{\mathrm{d}x}\right)^2\right]^{3/2}}{\dfrac{\mathrm{d}^2 y}{\mathrm{d}x^2}} = \dfrac{(\dot{x}^2+\dot{y}^2)^{\frac{3}{2}}}{\dot{x}\ddot{y}-\dot{y}\ddot{x}}$

Mathematical Constants and Formulae (Cont.)

Mensuration

Area of triangle, (sides a, b, c): $\triangle = \frac{1}{2}bc \sin A$
or $\sqrt{s(s-a)(s-b)(s-c)}$ where $2s = a+b+c$

Circle (radius r): Perimeter $= 2\pi r$ Area $= \pi r^2$

Ellipse (axes $2a$, $2b$): Perimeter $\approx 2\pi \sqrt{\dfrac{a^2+b^2}{2}}$ Area $= \pi ab$

Cylinder (radius r, height h): Area $= 2\pi r(h+r)$, Volume $= \pi r^2 h$

Area of curved surface of cone $= \pi rl$, where $l =$ slant height

Volume of cone or pyramid $= \frac{1}{3}Ah$, where $A =$ base area, $h =$ height

Sphere (radius r): Area $4\pi r^2$, Volume $(\frac{4}{3})\pi r^3$

Area cut off on sphere by parallel planes h apart $= 2\pi rh$

Trigonometry

(a) $\sin(\theta \pm \varphi) = \sin\theta\cos\varphi \pm \cos\theta\sin\varphi$

$\cos(\theta \pm \varphi) = \cos\theta\cos\varphi \mp \sin\theta\sin\varphi$

$\tan(\theta \pm \varphi) = \dfrac{\tan\theta \pm \tan\varphi}{1 \mp \tan\theta\tan\varphi}$

$\sin 2\theta = 2\sin\theta\cos\theta$

$\cos 2\theta = \cos^2\theta - \sin^2\theta = 2\cos^2\theta - 1 = 1 - 2\sin^2\theta$

$\sin 3\theta = 3\sin\theta - 4\sin^3\theta$, $\cos 3\theta = 4\cos^3\theta - 3\cos\theta$

$\sin A + \sin B = 2\sin\frac{1}{2}(A+B)\cos\frac{1}{2}(A-B)$

$\sin A - \sin B = 2\cos\frac{1}{2}(A+B)\sin\frac{1}{2}(A-B)$

$\cos A + \cos B = 2\cos\frac{1}{2}(A+B)\cos\frac{1}{2}(A-B)$

$\cos A - \cos B = -2\sin\frac{1}{2}(A+B)\sin\frac{1}{2}(A-B)$

If $\tan\frac{1}{2}x = t$, $\sin x = \dfrac{2t}{1+t^2}$ $\cos x = \dfrac{1-t^2}{1+t^2}$

$\tan x = \dfrac{2t}{1-t^2}$ $dx = \dfrac{2}{1+t^2}\,dt$

(b) In any triangle: $\dfrac{a}{\sin A} = \dfrac{b}{\sin B} = \dfrac{c}{\sin C} = 2R$ (sine rule)

$a^2 = b^2 + c^2 - 2bc\cos A$ (cosine rule)

$\sin\dfrac{A}{2} = \sqrt{\dfrac{(s-b)(s-c)}{bc}}$ $\cos\dfrac{A}{2} = \sqrt{\dfrac{s(s-a)}{bc}}$

Radius of circumcircle, $R = abc/4\triangle$ (where $\triangle =$

Radius of inscribed circle, $r = \dfrac{}{s}$ area of triangle)

Geometry

The polar form for a conic with origin at the focus is
$l = r(1 + e\cos\theta)$ For ellipse/hyperbola $e \lessgtr 1$

and foci are $(\pm ae, 0)$, directrices $x = \pm\dfrac{a}{e}$, where $b^2 = a^2|1-e^2|$

Solid angle: The solid angle of a cone is given by the
area intercepted by the cone on the surface of a
sphere of *unit* radius, with centre at the vertex.

When making measurements of a physical quantity, the final result is expressed as a number followed by the unit. The number expresses the ratio of the measured quantity to some fixed standard and the unit is the name or symbol for the standard. Over the years, a large number of standards have been defined for physical measurement and many systems of units have evolved. Recently, there has been an attempt to simplify the language of science by the adoption of a system of units, the Système Internationale d'Unités, which is intended to be used universally. This system of units, SI, was the outcome of a resolution of the 9th General Conference of Weights and Measures (CGPM) in 1948, which instructed an international committee to 'study the establishment of a complete set of rules for units of measurement.' The constants in this book are given in SI except where stated otherwise.

SI contains three classes of units: (*i*) base units, (*ii*) derived units, and (*iii*) supplementary units.

Base Units in SI: There are seven base units:
(*i*)   the metre, the standard of length,
(*ii*) the kilogram, the standard of mass,
(*iii*) the second, the standard of time,
(*iv*) the ampere, the standard of electric current,
(*v*)   the kelvin, the standard of temperature,
(*vi*) the candela, the standard of luminous intensity, and
(*vii*) the mole, the standard of amount of substance.

Derived Units: Derived units can be formed by combining base units. Thus the unit of force can be produced by combining the first three base units. Often derived units are given names, e.g. the unit of force is the *newton*.

Supplementary Units: Two supplementary units are at present defined, the radian and the steradian, which are the units for plane and solid angles respectively.

SI Prefixes and Multiplication Factors: To obtain multiples and submultiples of units, standard prefixes are used as shown below:

Multiplication factor	Prefix	Symbol
$1\ 000\ 000\ 000\ 000 = 10^{12}$	tera	T
$1\ 000\ 000\ 000 = 10^{9}$	giga	G
$1\ 000\ 000 = 10^{6}$	mega	M
$1\ 000 = 10^{3}$	kilo	k
$100 = 10^{2}$	hecto	h
$10 = 10^{1}$	deca	da
$0{\cdot}1 = 10^{-1}$	deci	d
$0{\cdot}01 = 10^{-2}$	centi	c
$0{\cdot}001 = 10^{-3}$	milli	m
$0{\cdot}000\ 001 = 10^{-6}$	micro	μ
$0{\cdot}000\ 000\ 001 = 10^{-9}$	nano	n
$0{\cdot}000\ 000\ 000\ 001 = 10^{-12}$	pico	p
$0{\cdot}000\ 000\ 000\ 000\ 001 = 10^{-15}$	femto	f
$0{\cdot}000\ 000\ 000\ 000\ 000\ 001 = 10^{-18}$	atto	a

It should be noted that masses are still expressed as multiples of the gram, although the base unit is the kilogram. Thus 10^{-6} kg should be written as 1 mg.

Other systems of units. Some other systems of units are still in common use. Thus for mechanical measurements, the British or fps system is still largely used, while for electrical measurements, the electrostatic and electromagnetic cgs systems are by no means obsolete. In the pages which follow, these systems of units are discussed and tables are included to help in conversion from one system to another.

2 The Fundamental Mechanical Units

(a) SI UNITS

In any system of measurement in mechanics, three fundamental units are required. These are the units of length, mass and time. The base units as used in SI are the metre, kilogram and second.

The metre (m)
This is defined as 1 650 763·73 of the wavelength, in vacuo of the orange light emitted by $^{86}_{36}$Kr in the transition $2p_{10}$ to $5d_5$.

The kilogram (kg)
This is defined as the mass of a platinum-iridium cylinder kept at Sèvres. Originally intended to be the mass of a cubic decimetre of water at its maximum density, the cylinder was subsequently found to be 28 parts per million too large. The cylinder was then taken as an arbitrary standard of mass, while the volume of water which had the same mass (at maximum density) was defined to be one litre (l). Thus 1 litre = 1000·028 cm³. The 1964 General Conference on Weights and Measures redefined the litre to be a cubic decimetre, but recommended that this unit should not be used in work of high precision.

The second (s)
This is the time taken by 9 192 631 770 cycles of the radiation from the hyperfine transition in caesium when unperturbed by external fields. Alternatively the *ephemeris* second is defined as 1/31 556 925·974 7 of the tropical year for 1900.

Derived units of length, mass and time
Through common usage, certain multiples and submultiples of the three fundamental units have been given names. A list of the more common ones is given below as they have been in frequent use. None of them is a recognised SI unit.

Length and area	Mass	Time
Micron $(\mu m) = 10^{-6}$ m	Tonne (t) $= 10^6$ g	Minute (min) $= 60$ s
Angstrom (Å) $= 10^{-10}$ m	$= 1000$ kg	Hour (h) $= 3\,600$ s
Fermi (fm) $= 10^{-15}$ m		Day (d) $= 86\,400$ s
Are (a) $= 100$ m²		Year (a) $\simeq 3\cdot1557 \times 10^7$ s
Barn (b) $= 10^{-28}$ m²		

SUPPLEMENTARY UNITS

The radian (rad) is the plane angle between two radii of a circle which cut off on the circumference an arc equal in length to the radius.

The steradian (sr) is the solid angle which, having its vertex in the centre of a sphere, cuts off an area of the surface of the sphere equal to that of a square with sides of length equal to the radius of the sphere.

Other units of angular measure are:
   The degree (°) is a unit of angle equal to $(\pi/180)$ rad.
   The minute of arc (') is $(1/60)$ degree and thus is equal to $(\pi/10\ 800)$ rad.
   The second of arc (") is $(1/60)$ minute and thus is equal to $(\pi/648\ 000)$ rad.

(b) THE CGS SYSTEM

A lot of early scientific work was done using the centimetre, gram and second as the base units in mechanics. Derived units in the cgs system were given names and some of them are still used. The table below lists the more common of the named derived units in SI and cgs with the conversion factors. The International Union of Pure and Applied Physics has recommended that certain symbols be used in scientific work and these are also included in the first column.

DERIVED UNITS IN SI AND CGS

Quantity and recommended symbol	Dimensions	SI unit	cgs unit	Ratio cgs/SI units
Mass m	M	kilogram (kg)	gram (g)	10^{-3}
Length l	L	metre (m)	centimetre (cm)	10^{-2}
Time t	T	second (s)	second (s)	1
Area A, S	L^2	m^2	cm^2	10^{-4}
Volume V	L^3	m^3	cm^3	10^{-6}
Density ρ	ML^{-3}	$kg\ m^{-3}$	$g\ cm^{-3}$	10^3
Velocity u, v	LT^{-1}	$m\ s^{-1}$	$cm\ s^{-1}$	10^{-2}
Acceleration a	LT^{-2}	$m\ s^{-2}$	gal	10^{-2}
Momentum p	MLT^{-1}	$kg\ m\ s^{-1}$	$g\ cm\ s^{-1}$	10^{-5}
Moment of Inertia I, J	ML^2	$kg\ m^2$	$g\ cm^2$	10^{-7}
Angular Momentum L	ML^2T^{-1}	$kg\ m^2\ s^{-1}$	$g\ cm^2\ s^{-1}$	10^{-7}
Force F	MLT^{-2}	newton (N)	dyne (dyn)	10^{-5}
Energy or Work E, W	ML^2T^{-2}	joule (J)	erg	10^{-7}
Power P	ML^2T^{-3}	watt (W)	$erg\ s^{-1}$	10^{-7}
Pressure or Stress p	$ML^{-1}T^{-2}$	pascal (Pa)	$dyn\ cm^{-2}$	10^{-1}
Surface Tension γ	MT^{-2}	$N\ m^{-1}$	$dyn\ cm^{-1}$	10^{-3}
Viscosity η	$ML^{-1}T^{-1}$	$kg\ m^{-1}\ s^{-1}$	poise	10^{-1}
Frequency ν, f	T^{-1}	hertz (Hz)	s^{-1}	1

NOTE: The ratio in the final column is that of the *actual units*. Thus the SI unit of pressure, the pascal, is 10 times larger than the cgs unit, the dyn cm^{-2}. This means that a pressure of 1 pascal is the same as a pressure of 10 dyn cm^{-2}.

(c) THE BRITISH OR fps SYSTEM

In this system of units, the standards of length and mass are the *foot* and the *pound*. The unit of time is the *second* which is defined as in the metric system.

The foot (ft)
The foot is one-third of the Imperial Standard yard which is now defined to be 0·9144 metre exactly. Thus the foot is defined as 0·3048 metre exactly.

The pound (lb)
This is now defined to be 0·453 592 37 kilogram exactly.

The gallon (gal)
This unit of volume is also defined in the British system. It is the volume occupied by exactly 10 pounds of water of density 0·988 859 gram per millilitre weighed in air of density 0·001 217 gram per millilitre against weights of density 8·136 grams per millilitre.

SECONDARY UNITS IN THE BRITISH SYSTEM
The following list shows the most common secondary units in the British system.

Units of Length

12 inches	= 1 foot (ft)
3 feet	= 1 yard (yd)
22 yards	= 1 chain
10 chains	= 1 furlong
8 furlongs or 1760 yards	= 1 mile (mi)
6080 feet	= 1 UK nautical mile*
6 feet	= 1 fathom

Units of Mass

16 ounces (oz)	= 1 pound (lb)
14 pounds (lb)	= 1 stone
28 pounds	= 1 quarter
4 quarters or 112 pounds	= 1 hundredweight
20 hundredweight (cwt) or 2240 lb	= 1 ton

Units of Area

4840 square yards	= 1 acre
640 acres	= 1 square mile

Units of Volume

20 fluid ounces (fl. oz)	= 1 pint
2 pints (pt)	= 1 quart
4 quarts (qt)	= 1 gallon

*The Nautical mile is the average distance on the earth's surface subtended by one minute of latitude. The UK nautical mile is 6080 ft but the International nautical mile, which is used by the Admiralty and most other nations, measures 1852 m.

fps unit	SI unit	Reciprocal
length 1 inch (in)	$= 2\cdot54 \times 10^{-2}$ m	$39\cdot370\ 079$
1 foot (ft)	$= 0\cdot3048$ m	$3\cdot280\ 839$
1 yard (yd)	$= 0\cdot9144$ m	$1\cdot093\ 613$
1 fathom	$= 1\cdot828\ 8$ m	$0\cdot546\ 806$
1 chain	$= 20\cdot116\ 8$ m	$4\cdot970\ 970 \times 10^{-2}$
1 furlong	$= 2\cdot011\ 68 \times 10^2$ m	$4\cdot970\ 970 \times 10^{-3}$
1 mile (mi)	$= 1\cdot609\ 344 \times 10^3$ m	$6\cdot213\ 712 \times 10^{-4}$
Area 1 in^2	$= 6\cdot451\ 6 \times 10^{-4}$ m^2	$1\cdot550\ 003 \times 10^3$
1 ft^2	$= 9\cdot290\ 304 \times 10^{-2}$ m^2	$10\cdot763\ 910$
1 yd^2	$= 0\cdot836\ 127$ m^2	$1\cdot195\ 990$
1 mi^2	$= 2\cdot589\ 988 \times 10^6$ m^2	$3\cdot861\ 022 \times 10^{-7}$
1 acre	$= 4\cdot046\ 856 \times 10^3$ m^2	$2\cdot471\ 054 \times 10^{-4}$
Volume 1 in^3	$= 1\cdot638\ 706 \times 10^{-5}$ m^3	$6\cdot102\ 374 \times 10^4$
1 ft^3	$= 2\cdot831\ 685 \times 10^{-2}$ m^3	$35\cdot314\ 67$
1 yd^3	$= 0\cdot764\ 555$ m^3	$1\cdot307\ 950$
1 fluid ounce (fl oz)	$= 2\cdot841\ 306 \times 10^{-5}$ m^3	$3\cdot519\ 508 \times 10^4$
1 pint (pt)	$= 5\cdot682\ 613 \times 10^{-4}$ m^3	$1\cdot759\ 754 \times 10^3$
1 quart (qt)	$= 1\cdot136\ 523 \times 10^{-3}$ m^3	$8\cdot798\ 770 \times 10^2$
1 gallon (gal)	$= 4\cdot546\ 09 \times 10^{-3}$ m^3	$2\cdot199\ 692 \times 10^2$
1 bushel (bu)	$= 0\cdot036\ 369$ m^3	$27\cdot495\ 944$
1 gallon USA (= 231 in^3)	$= 3\cdot785\ 412 \times 10^{-3}$ m^3	$2\cdot641\ 721 \times 10^2$
Mass 1 ounce (oz)	$= 2\cdot834\ 952 \times 10^{-2}$ kg	$35\cdot273\ 962$
1 pound (lb)	$= 0\cdot453\ 592\ 37$ kg	$2\cdot204\ 623$
1 stone	$= 6\cdot350\ 293$ kg	$0\cdot157\ 473$
1 quarter	$= 12\cdot700\ 586$ kg	$7\cdot873\ 652 \times 10^{-2}$
1 hundredweight (cwt)	$= 50\cdot802\ 345$ kg	$1\cdot968\ 413 \times 10^{-2}$
1 ton	$= 1\cdot016\ 047 \times 10^3$ kg	$9\cdot842\ 065 \times 10^{-4}$
Density 1 lb ft^{-3}	$= 16\cdot018\ 463$ kg m^{-3}	$6\cdot242\ 796 \times 10^{-2}$
Speed 1 in s^{-1}	$= 2\cdot54 \times 10^{-2}$ m s^{-1}	$39\cdot370\ 079$
1 ft s^{-1}	$= 0\cdot3048$ m s^{-1}	$3\cdot280\ 839$
1 mi h^{-1}	$= 0\cdot447\ 04$ m s^{-1}	$2\cdot236\ 936$
Force 1 poundal (pdl)	$= 0\cdot138\ 255$ N	$7\cdot233\ 011$
1 lbf (i.e. the wt of 1 lb mass)	$= 4\cdot448\ 222$ N	$0\cdot224\ 809$
Pressure 1 lbf in^{-2} (p.s.i.)	$= 6\cdot894\ 757 \times 10^3$ Pa	$1\cdot450\ 377 \times 10^{-4}$
Energy 1 ft pdl	$= 4\cdot214\ 011 \times 10^{-2}$ J	$23\cdot730\ 360$
1 ft lbf	$= 1\cdot355\ 817$ J	$0\cdot737\ 562$
1 Btu	$= 1\cdot055\ 06 \times 10^3$ J	$9\cdot478\ 134 \times 10^{-4}$
1 therm	$= 1\cdot055\ 06 \times 10^8$ J	$9\cdot478\ 134 \times 10^{-9}$
Power 1 horse power (hp)	$= 7\cdot457\ 00 \times 10^2$ W	$1\cdot341\ 022 \times 10^{-3}$
Standard atmosphere	$14\cdot695\ 916$ lbf in$^{-2} = 1\cdot013\ 25 \times 10^5$ Pa	
Standard acceleration of gravity	$32\cdot174\ 05$ ft s$^{-2} = 9\cdot806\ 65$ m s^{-2}	

When units were first required for the measurement of electrical quantities it was natural to define them in terms of the three fundamental units, centimetre, gram and second, which were already commonly used in mechanics. Electrical phenomena are related to mechanical phenomena by two effects: (a) the force between static electric charges (Coulomb's law) and (b) the force between electric currents (Ampere's law). Correspondingly, two distinct systems of cgs electrical units were introduced: the electrostatic and electromagnetic systems.

Neither of these systems has units of convenient size in practical applications. Consequently, a practical set of electrical units, defined as decimal multiples of the electromagnetic units was established by various International Congresses of Electricians meeting between 1881 and 1889. The first two units defined were the ohm (10^9 emu), chosen to be similar to the Siemens unit of resistance, and the volt (10^8 emu), chosen to be similar to the emf of the Daniell cell. From these, six other units, the ampere, coulomb, joule, watt, henry and farad were derived. These practical units were not made into a complete system, since no magnetic units were defined, the unmodified magnetic units of the electromagnetic system (e.g. oersted and gauss) being used in practical applications.

In 1901, Giorgi showed that if the metre, kilogram, and second were used as fundamental units instead of the centimetre, gram and second, a single, consistent and comprehensive system of electrical and magnetic units could be built up incorporating the already firmly-established practical units. This is because, using the metre, kilogram and second, the unit of mechanical power becomes 10^7 erg s^{-1}, which is the appropriate practical electrical unit, i.e. the watt. The use of the Giorgi system, also known as the mks system, or the Absolute Practical System was approved by an International Electro-technical Commission in 1935. The Absolute Practical System, with the ampere as the electrical base unit was adopted by the CGPM for SI.

RELATIONS BETWEEN THE SYSTEMS OF ELECTRICAL UNITS

Coulomb's law for the force F between charges Q_1 and Q_2, distance r apart in vacuo, may be expressed in the form

$$F = \frac{Q_1 Q_2}{\varepsilon_i r^2} \tag{1}$$

where ε_i is a constant, called the permittivity of free space. In the cgs electrostatic system, ε_i is chosen to be unity. This choice of the value of ε_i, together with the use of the centimetre, gram and second uniquely determines the system of units.

Ampere's law for the force between two parallel current elements $I_1 ds_1$ and $I_2 ds_2$, distance r apart in vacuo, may be expressed in the form

$$F = \mu_i \frac{I_1 ds_1 I_2 ds_2 \sin \theta}{r^2} \tag{2}$$

where μ_i is a constant, called the permeability of free space. In the cgs electromagnetic system μ_i is chosen to be unity. This choice of the value of μ_i, together with the use of the centimetre, gram and second, again determines the system of units uniquely.

It may be noted that these two systems of units, defined by $\varepsilon_i = 1$ and $\mu_i = 1$, cannot be combined directly to form a single consistent system. It can be shown from Maxwell's electromagnetic theory that, in any consistent system of units, $\mu_i \varepsilon_i = 1/c^2$, where c is the velocity of electromagnetic radiation (e.g. light) in free space, measured in the appropriate units of length and time (e.g. $c \simeq 3 \times 10^8$ m s^{-1}).

In SI, neither ε_i nor μ_i is chosen to be unity. The fundamental units chosen are the metre, kilogram, and second and ampere which are sufficient to determine the complete system uniquely. In particular, μ_i may be shown to have the value 10^{-7} newton ampere^{-2}, where the newton is the SI unit of force. This value of μ_i is readily derived from equation (2). The appropriate value of ε_i is then calculated, knowing the experimentally determined value of the velocity of light.

RATIONALIZATION OF MKS UNITS

It is found that many formulae are simplified if the permeability of free space is re-defined as $\mu_0 = 4\pi\mu_i$. Ampere's law for current elements in free space is then expressed in 'Rationalised mks units' as

$$F = \frac{\mu_0 I_1 ds_1 I_2 ds_2 \sin\theta}{4\pi r^2} \tag{3}$$

where $\mu_0 = 4\pi \times 10^{-7}$ newton ampere^{-2} (or henry metre^{-1}).

Similarly, the permittivity of free space is re-defined as $\varepsilon_0 = \varepsilon_i/4\pi$, and Coulomb's law, for charges in free space, is expressed in rationalised mks units as

$$F = \frac{Q_1 Q_2}{4\pi \varepsilon_0 r^2} \tag{4}$$

The value of ε_0, given by $1/c^2\mu_0$, is approximately $8 \cdot 85 \times 10^{-12}$ farad metre.$^{-1}$

For an isotropic, homogeneous medium other than free space, μ_0 in equation (3) is replaced by $\mu = \mu_r\mu_0$, where μ_r is the relative permeability of the medium; and ε_0 in equation (4) is replaced by $\varepsilon = \varepsilon_r\varepsilon_0$, where ε_r is the relative permittivity (dielectric coefficient) of the medium.

DEFINITIONS OF ELECTRIC AND MAGNETIC QUANTITIES IN SI

The base unit

Current (I): The unit of current is the *ampere* (A), defined as that constant current which, if maintained in each of two infinitely long straight parallel wires of negligible cross-section placed 1 metre apart, in a vacuum, will produce between the wires a force of 2×10^{-7} newtons per metre length.

Derived units

Charge (Q): The unit of charge (quantity) is the *coulomb* (C), defined as the quantity of electricity transported per second by a current of 1 ampere.

Potential Difference (V): The unit of potential difference is the *volt* (V), defined as that difference of electrical potential between two points of a wire carrying a constant current of 1 ampere when the power dissipation between those points is 1 watt.

Resistance (R): The unit of resistance is the *ohm* (Ω), defined as the electrical resistance between two points of a conductor when a constant potential difference of 1 volt applied between these points produces in the conductor a current of 1 ampere.

Conductance (G): The unit of conductance is the *siemens*, (S) defined as the electrical conductance between two points of a conductor when a constant potential difference of 1 volt applied between these points produces in the conductor a current of 1 ampere.

Inductance (L): The unit of inductance is the *henry*, (H) defined as the inductance of a closed circuit in which an electromotive force of 1 volt is produced when the current in the circuit varies uniformly at the rate of 1 ampere per second.

Capacitance (C): The unit of capacitance is the *farad*. (F) defined as the capacitance of a capacitor between the plates of which there appears a difference of potential of 1 volt when it is charged by 1 coulomb.

Magnetic Intensity (H): is defined through Ampere's theorem for the intensity due to a current element. In the usual notation $H = \dfrac{I.ds.\sin \theta}{4\pi r^2}$. Unit *ampere metre^{-1}*.

Magnetic Flux (Φ) **of the induction B**: is defined as $\int \mathbf{B.n} \, dA$ where **n** is a unit vector perpendicular to an element of area dA. Unit, *weber* (Wb).

Magnetic Flux Density or Induction (B): is defined through the equation for the force on a current element placed in a magnetic field, viz $F = B.I.ds.\sin \theta$, in the usual notation. Unit, Tesla (T). $B = \mu_0\mu_r H$ where μ_r is the relative permeability of the medium with respect to free space and μ_0 is the permeability of free space. $\mu_0 = 4\pi \times 10^{-7}$ henry metre^{-1}.

Magnetic Moment (m)*: is the couple exerted on a magnet placed at right angles to a uniform field with unit flux density. Unit, *ampere metre^2*.

Intensity of Magnetisation (M)*: is the magnetic moment per unit volume of a magnet. Unit, *ampere metre^{-1}*.

Pole Strength (P)*: On the mks system the hypothetical concept of an isolated magnetic pole is abandoned by many writers on the grounds that all magnetism arises from electrical effects, hence the definitions of H and B (above). Other writers use the idea of a magnetic pole as a simple and convenient concept in magnetometry. In this connection we define a unit magnetic pole as one which when situated 1 metre distant in vacuum from an equal pole experiences a force of $\mu_0/4\pi$ newtons. Alternatively it can be defined as that pole strength which when placed in a unit induction experiences a force of 1 newton. Unit, *ampere-metre*.

Line of Force: A line of force is a curve in a magnetic field, such that the tangent at every point is the direction of the magnetic force at that point.

Magnetomotive Force (F_m): is defined as the line integral $\int H . dl$ evaluated for a closed path. It is equal to the total conduction current linked. Unit, *ampere*.

Coulomb's Magnetic Law: states that the force between two poles P_1 and P_2 situated distance d apart is given by $F = \dfrac{\mu_r\mu_0 P_1 P_2}{4\pi d^2}$, where μ_r is the permeability of the medium and μ_0 the permeability of the free space $= 4\pi \times 10^{-7}$ henry metre^{-1}.

$\mu_r\mu_0$ replaces the permeability μ of the cgs system.

Electrical Intensity $(X$ or $E)$: The electrical intensity at a point in an electric field is the force exerted on unit charge (1 coulomb) placed at that point, assuming that the field is not disturbed by so doing. Unit, *volt metre^{-1}* (which is equivalent to the newton coulomb^{-1}).

Coulomb's Electrostatic Law: appears in the form $F = \dfrac{Q_1 Q_2}{4\pi\varepsilon_0\varepsilon_r d^2}$, where Q_1 and Q_2 are the two charges situated a distance d apart in a medium whose permittivity relative to that of free space is ε_r. The permittivity of free space $\varepsilon_0 \simeq (1/36\pi) \times 10^{-9}$ farad metre^{-1}. ε_r is a pure number. (The product $\varepsilon_0\varepsilon_r$ is analogous to the dielectric constant as defined in the cgs system.)

---

*For these definitions we adopt the Sommerfeld system of units in which the magnetic moment of a current loop is the product of the area of the loop and the current flowing round the edge of the loop: $m = IA$. An alternative system due to Kennelly uses the relation $m = \mu_0 IA$.

DEFINITIONS IN THE CGS ELECTROMAGNETIC SYSTEM OF UNITS

Magnetic Pole (P): When two like magnetic poles are placed 1 cm apart in vacuo, they repel one another with a force of 1 dyne.

Magnetic Field Strength or Intensity (H): is the force experienced by a unit North pole when placed at the given point in a magnetic field, it being assumed that the introduction of the pole does not disturb the field. Unit, *oersted*. The intensity is one oersted when a unit North pole experiences a force of 1 dyne on being placed at the given point in the field. The field strength in vacuum is represented as the number of lines of force passing perpendicularly through 1 cm² placed at the point in question. On this convention 4π lines of force leave a unit North pole.

Magnetic Flux (Φ): through any area at right angles to a magnetic field is the product of the area and the field strength. Unit, *maxwell*. One maxwell is the flux through unit area (1 cm²) placed perpendicularly to a unit uniform field. Hence one line of force is equivalent to one maxwell.

Magnetic Moment (m): of a magnet, is the couple exerted on the magnet when placed at right angles to a unit uniform field. For a bar magnet it is equivalent to the product of the pole strength and the distance between the poles. Unit, *pole cm*.

Magnetic Potential (Ω): is the work done in bringing a unit North pole from infinity or a point of zero potential to the point in question. Unit, *gilbert*. 1 gilbert is that potential which requires the expenditure of 1 erg of work in bringing a unit North pole from infinity to the point.

Intensity of Magnetisation (M): of a sample of material is the magnetic moment per unit volume.

Magnetic Susceptibility (χ): of a material is the ratio of the intensity of magnetisation produced in the sample to the magnetic field which produced the magnetisation. $\chi = \dfrac{M}{H}$.

N.B.: χ is not a constant but is a function of H.

Magnetic Induction (B): in any material is the number of lines of magnetic force (often called lines of induction) passing perpendicularly through unit area. Unit, *gauss*. One gamma = 10^{-5} gauss.

Magnetic Permeability (μ): of any material is the ratio of the magnetic induction in the sample to the magnetic field producing it, i.e. $\mu = B/H$. Although μ is so defined, B is not proportional to H, for $B = H + 4\pi M$. Also $B/H = 1 + 4\pi M/H$ or $\mu = 1 + 4\pi\chi$. Hence μ is not a constant but a function of H. (see χ above)

Coulomb's Law of Force: states that the force F between two poles of strength P_1 and P_2 is proportional to the product of the pole strengths and inversely proportional to the square of their distance apart (d). Thus $F = \dfrac{P_1 P_2}{\mu d^2}$ where $1/\mu$ is the constant of proportionality, μ being the permeability of the medium in which the poles are located. In this system, as already stated, the permeability of free space is defined to be unity.

Current (I): The electromagnetic unit (emu) of current is that which when flowing round 1 cm arc of a circle of radius 1 cm, produces a magnetic field of 1 oersted at the centre. Unit, *emu of current*.

Charge (Q): The emu of charge (quantity) is that delivered in 1 second by the passage of unit current. Unit, *biot*.

Potential Difference (P.D.): When unit current flows between two points in a circuit and unit work (1 erg) is done per second, the P.D. between the two points is unity. Unit, *emu of P.D.*

Electromotive Force (emf): When lines of magnetic force cut a conductor an emf is created which is given numerically (in emu) by the number of lines cut per second. Emf $= dn/dt$.

Resistance (R): A conductor has unit resistance when on applying unit P.D., unit current flows. Unit, *emu of resistance*.

Self Inductance (L): A conductor possesses unit self inductance if unit emf is developed across it when the rate of change of current is unity. Unit, *emu of self inductance*.

Mutual Inductance (M): Two conductors possess unit mutual inductance when unit emf is developed in one by unit rate of change of current in the other. Unit, *emu of mutual inductance*.

DEFINITIONS IN THE CGS ELECTROSTATIC SYSTEM OF UNITS

Electric Charge (Q): When two like unit electrical charges are placed 1 cm apart in vacuum, they repel one another with a force of 1 dyne. Unit, *franklin*.

Electric Field Strength (Intensity) (E): The electric field at a point has unit strength if a unit positive charge experiences a force of 1 dyne when placed at the point, it being assumed that the introduction of the charge does not disturb the field. Unit, *dyne per franklin*.

Electrical Potential (V): is the work done in conveying a unit positive charge from infinity or a point of zero potential to the point in question against the forces of the field. Unit, *erg per franklin*.

Capacitance (*C*): A conductor has unit capacitance when the addition of unit charge raises its potential by unity. Unit, *cm*.

Dielectric Constant or Specific Inductivity Capacity (ε_r): of a material is the ratio of the capacity of a condenser with the material as dielectric to that of the same condenser in vacuum without a material dielectric.

Coulomb's Electrostatic Law of Force: states that the force *F* between two charges Q_1 and Q_2 is proportional to the product of the charges and inversely proportional to the square of their distance apart *d*. Thus $F = \dfrac{Q_1 Q_2}{\varepsilon_r d^2}$, where $1/\varepsilon_r$ is the constant of proportionality. ε_r is the dielectric constant of the medium in which the charges are located. On the electrostatic system of units, the dielectric constant of free space is unity.

RELATIONS BETWEEN SI, AND cgs ELECTROSTATIC AND ELECTROMAGNETIC UNITS

Quantity and preferred symbol		SI unit and abbreviation		cgs units	
				esu	emu
Mass	*m*	1 kilogram	kg	10^3 gram	10^3 gram
Length	*l*	1 metre	m	10^2 cm	10^2 centimetre
Time	*t*	1 second	s	1 second	1 second
Current	*I*	1 ampere	A	10c	10^{-1} biot
Charge	*Q*	1 coulomb	C	10c franklin	10^{-1}
Potential difference	*V*	1 volt	V	$10^6/c$	10^8
Power	*P*	1 watt	W	10^7 erg s^{-1}	10^7 erg s^{-1}
Resistance	*R*	1 ohm	Ω	$10^5/c^2$	10^9
Conductance	*G*	1 siemens	S	$10^{-5}c^2$	10^{-9}
Inductance	*H*	1 henry	H	$10^5/c^2$	10^9
Capacitance	*F*	1 farad	F	$10^{-5}c^2$	10^{-9}
Magnetic flux	*Φ*	1 weber	Wb	$10^6/c$	10^8 maxwell
Magnetic induction	*B*	1 tesla	T	$10^2/c$	10^4 gauss
*Magnetic field strength	*H*	1 ampere metre^{-1}	A m^{-1}	$4\pi c/10$	$4\pi \times 10^{-3}$ œrsted
*Magnetization	*M*	1 ampere metre^{-1}	Am^{-1}	$10^{-1}c$	10^{-3}
Electric field strength	*E*	1 volt metre^{-1}	V m^{-1}	$10^4/c$	10^6
Electric displacement	*D*	1 coulomb metre^{-2}	C m^{-2}	$4\pi \times 10^{-3}c$	$4\pi \times 10^{-5}$

NOTE: in the table, c represents the speed of light in vacuo. In SI units, $c \simeq 3 \times 10^8$ m s^{-1}.

*The apparent discrepancy in the conversions of magnetization and magnetic field strength arises from the different definitions of magnetization. Following Sommerfeld, magnetization is now defined by the equation $B = \mu_0(H + M)$. In cgs, it was defined by the equation, $B = H + 4\pi M$.
N.B.—For an exhaustive account of systems of electrical units see: L. Young, 'Systems of units in electricity and magnetism'. (Edinburgh, Oliver and Boyd, 1969).

It is frequently important to express an amount of substance in terms of a fixed number of constituent parts. This has been done by referring to gram-atom or gram-molecule of a substance. In SI, the amount of a substance is expressed relative to a fixed mass of the isotope of carbon containing 6 protons and 6 neutrons in its nucleus, $^{12}_{6}C$. It is possible to measure the atomic masses of other substances in units of the mass of $^{12}_{6}C$ very accurately.

SI base unit, the Mole (mol)

The mole is the amount of substance of a system which contains as many elementary entities as there are atoms in 0·012 kilogram of the carbon isotope $^{12}_{6}C$. Note: When the mole is used, it is essential to specify the elementary entities under consideration. These may be atoms, molecules, ions, electrons or other particles or groups of particles.

The unified mass unit (u)

As it is possible to measure atomic masses relative to the mass of $^{12}_{6}C$ with extreme accuracy, it is useful to have a mass scale based on the mass of this atom. On the unified mass scale, the mass of the nuclide $^{12}_{6}C$ is set to be exactly 12·0 u.

The chemical and physical mass scales

In past chemical practice, atomic weights were often expressed on the chemical mass scale in which the atomic weight of naturally occurring oxygen was taken to be exactly 16·0. In view of the uncertainty of the isotopic composition of oxygen, another scale, the physical mass scale, came into use. On this scale, the mass of the isotope $^{16}_{8}O$ was set to be exactly 16·0. The IUPAP and the IUCAC jointly agreed in 1959/60 that these scales be discontinued and the unified mass scale used instead.

1 unified mass unit ($^{12}C = 12$) = 1·000 317 92 physical mass unit ($^{16}O = 16$)

1 unified mass unit = $1·660\ 43 \times 10^{-27}$ kg

1 chemical mass unit = $1·660\ 24 \times 10^{-27}$ kg

The Avogadro constant

Avogadro's law states that under the same conditions of temperature and pressure, equal volumes of all gases contain equal numbers of molecules. Avogadro's number was then defined as the number of entities in a gram-atom or gram-molecule of a substance. Different values of this number were then needed, depending on the mass scale used. In SI, the Avogadro constant is defined as the number of atoms in 0·012 kg of the isotope, $^{12}_{6}C$, and is thus the number of entities in a mole of substance.

5 Heat Units and Definitions

Temperature (t, θ or T). In SI, temperatures are measured on the thermodynamic scale with the Absolute Zero as zero and the triple point of water (i.e. the temperature at which ice, water and water vapour are in equilibrium) as the upper fixed point. The thermodynamic scale is that given by a theoretical Carnot heat engine and is equal to the perfect gas scale.

SI base unit, the Kelvin (K). The kelvin (K) unit of thermodynamic temperature, is the fraction $1/273 \cdot 16$ of the thermodynamic temperature of the triple point of water.

The Degree Celsius (°C). The centigrade scale of temperature used the ice point as zero and the boiling point of water at 1 standard atmosphere as the upper fixed point set to be 100°C. The Celsius scale of temperature is defined to be the same as the thermodynamic scale with the zero shifted to the ice point, which is $273 \cdot 15$ K, and thus:

$$\theta/°C = T/K - 273 \cdot 15$$

The International Practical Scale of Temperature (IPST). In view of the difficulty of measuring on the thermodynamic scale, a scale of temperature based on fixed points was suggested by the 7th CCPM in 1927. The scale has been revised since so as to make temperatures on this scale agree as nearly as possible with the thermodynamic Celsius scale. A list of the fixed points and other important temperatures will be found on p. 70.

The Degree Fahrenheit (°F). On the Fahrenheit scale, the ice point is 32°F and the steam point, 212°F. Thus $t/°F = 32 + 1.8\,(\theta/°C)$.

The Degree Reaumur (°R). On the Reaumur scale, the ice point is 0°R and the steam point, 80°R. Thus $t/°R = 0.8\,(\theta/°C)$.

Quantity of Heat (Q). Quantities of heat are measured in joules (J) in SI. Other units have been used, notably the *calorie*. The calorie is the amount of heat required to raise the temperature of 1 gram of water by 1°C. This definition is not very precise however as the specific heat capacity of water varies with temperature. The *15° calorie* was defined as the heat required to raise the temperature of 1 g of water from $14 \cdot 5$ °C to $15 \cdot 5$°C. The *mean calorie* was defined as one hundredth of the heat required to raise the temperature of 1 g of water from 0°C to 100°C. The *kilocalorie* (1 000 calories) has also been used and written Calorie. Where quantities of heat are expressed in calories, it is recommended that the conversion factor to convert to joules be stated.

In the fps system, the *British thermal unit* (Btu) is used. This is the quantity of heat required to raise the temperature of 1 lb of water through 1°F. The therm is 10^5 Btu.

Specific Heat Capacity (c_p, c_v). This is the amount of heat required to raise the temperature of 1 kg of a substance 1 K. Units, $J\,kg^{-1}\,K^{-1}$.

Molar Heat Capacity (C_m). This is the amount of heat required to raise the temperature of 1 mol of substance through 1 K. Units, $J\,mol^{-1}\,K^{-1}$.

Heat Capacity (C). The amount of heat required to raise the temperature of a body through 1 K. Units, $J\,K^{-1}$.

Water Equivalent. The mass of water having the same total heat capacity as the given body.

Thermal Conductivity (λ). The rate of flow of heat (dQ/dt) through a surface of area, A, in a medium is given by:

$$\frac{dQ}{dt} = -\lambda A \frac{dT}{dx},$$

where (dT/dx) is the temperature gradient, measured in the direction normal to the surface. The quantity λ, is the thermal conductivity of the medium. Units, $J\ m^{-1}\ s^{-1}\ K^{-1}$, or $W\ m^{-1}\ K^{-1}$.

Specific Latent Heat (l). The specific latent heat of fusion (specific enthalpy change on fusion) of a body is the heat required to convert 1 kg of the solid at its melting point into liquid at the same temperature. Unit, $J\ kg^{-1}$.

The specific latent heat of vaporization (enthalpy change on vaporization) of a liquid is the heat required to convert 1 kg of the liquid at its boiling point into vapour at the same temperature. Unit, $J\ kg^{-1}$.

Linear Expansivity (α). The increase in length per unit length per unit rise in temperature. Unit, K^{-1}.

Cubic Expansivity (γ). The increase in volume per unit volume per unit rise in temperature. Unit, K^{-1}.

Critical Temperature (T_c) of a gas or vapour is that temperature above which it is not possible to liquefy the gas by the application of pressure alone. To liquefy a gas it must be cooled below its critical temperature before being compressed.

Critical Pressure (p_c): That pressure which just liquefies a gas at its critical temperature.

Critical Volume (V_c): The volume of unit mass of gas at its critical temperature and pressure, *i.e.* it is the reciprocal of the critical density. It is often taken as the volume of one mole of a gas at its critical temperature and pressure.

Radiation. Stefan–Boltzmann Law: The total energy, E, of all wavelengths radiated per second per square metre by a full radiator at temperature T to surroundings at T_0 is given by $E = \sigma(T^4 - T_0^4)$, where σ is Stefan's constant. $\sigma = 5 \cdot 669\ 7 \times 10^{-8}\ W\ m^{-2}\ K^{-4}$

Planck's Radiation Law: The energy density of radiation in an enclosure at temperature T having wavelengths in the range λ to $\lambda + d\lambda$ is $u_\lambda d\lambda$, where

$$u_\lambda d\lambda = 8\pi ch\lambda^{-5}(\exp hc/\lambda kT - 1)^{-1}\ d\lambda\ = c_1\lambda^{-5}(\exp c_2/\lambda T - 1)^{-1}d\lambda$$
$$c_1 = 4 \cdot 992\ 1 \times 10^{-24}\ J\ m \qquad c_2 = 1 \cdot 438\ 79 \times 10^{-2}\ m\ K$$

The corresponding relation for radiation of frequency, ν, is

$$u_\nu d\nu = (8\pi h/c^3)(\exp h\nu/kT - 1)^{-1}\nu^3 d\nu.$$

h = Planck's constant; c = speed of light; k = Boltzmann's constant; T = temperature of the enclosure.

Wien's Displacement Law: The wavelength of the most strongly emitted radiation in the continuous spectrum from a full radiator is inversely proportional to the absolute temperature of that body, *i.e.* $\lambda T = b$, where b is Wien's constant = $2 \cdot 898 \times 10^{-3}\ m\ K$.

The Energy (E) of a quantum of radiation of frequency ν is $E = h\nu$ where h is Planck's constant.

Luminous intensity. In SI, the unit of luminous intensity is the candela. This unit replaces the International candle which was defined in terms of the light emitted per second in all directions by a specified electric lamp.

SI base unit, the candela (cd). The candela is the luminous intensity, in the perpendicular direction, of a surface of 1/600 000 square metre of a full radiator at the temperature of freezing platinum under a pressure of 101 325 newtons per square metre.

$$1 \text{ candela} = 0\text{·}982 \text{ international candles.}$$

Luminous flux: The unit of luminous flux, the lumen (lm) is defined as the light energy emitted per second within unit solid angle by a uniform point source of unit luminous intensity. Thus $1 \text{ cd} = 1 \text{ lm sr}^{-1}$.

Illuminance of a surface is defined as the luminous flux reaching it perpendicularly per unit area. The British unit is the lumen ft^{-2}, formerly called the foot candle (f.c.). The metric unit is the lumen m^{-2} or lux (lx).

Lambert's Cosine law: For a surface receiving light obliquely, the illumination is proportional to the cosine of the angle which the light makes with the normal to the surface.

Brightness of a surface is that property by which the surface appears to emit more or less light in the direction of view. This is a subjective quantity. The corresponding physical measurement of the light actually emitted is called the luminance.

Luminance of a surface is the measure of the light actually emitted (*i.e.* the luminous intensity) per unit projected area of surface, the plane of projection being perpendicular to the direction of view. Unit, cd ft^{-2} or cd m^{-2}. In engineering, the luminance of an ideally diffusing surface emitting or reflecting one lumen ft^{-2} is called one foot-lambert (ft-L).

The Refractive index of a material (n) is the ratio of the velocity of light in free space to that in the material.

Snell's law: For light incident on a boundary between two media, the ratio of the sine of the angle of incidence (the angle between the light ray in the first medium and the normal to the boundary surface) to the sine of the angle of refraction (the angle between the refracted ray in the second medium and the normal) is a constant, being equal to the inverse ratio of the refractive indices of the two media.

Dioptre is the unit of measure of the power of a lens and is given numerically by the reciprocal of the focal length expressed in metres.

7 Acoustical Units and Definitions

Pressure: The unit of sound pressure is the pascal usually quoted as the root mean square (r.m.s.) pressure for a pure sinusoidal wave.

Frequency: The unit of frequency is the cycle per second, now designated the hertz (Hz).

Threshold of Hearing is, for a normal (average) observer, the sound level or intensity which is just audible. For a pure sinusoidal note of frequency 1000 Hz it is close to a root mean square pressure of 2×10^{-5} Pa.

Power Ratio: The unit of acoustical (or electrical) power measurement with respect to a standard level, is one *bel*. The interval between two powers W_1 and W_0 in bels is $\log_{10}(W_1/W_0)$. In practical work the decibel (dB) is used. The interval between two powers W_1 and W_0 is $10 \log_{10}(W_1/W_0)$ dB. In some instances it is more convenient to employ natural logarithms. The power ratio so obtained is called the neper and is defined as follows. The power interval between W_1 and W_0 is $\frac{1}{2} \log_e(W_1/W_0)$ nepers. Hence 1 neper = 8·686 dB.

Intensity (I) of a sound wave in a given direction is the sound energy transmitted per second in this direction through unit area placed perpendicularly to the specified direction. Unit, W m⁻². For a sinusoidal plane or spherical wave, the intensity is proportional to the mean square pressure exerted on an area at right angles to the given direction. Hence the interval between two intensities is given by $10 \log_{10}(I_1/I_0)$ dB or $20 \log_{10}(p_1/p_0)$ dB where p_1 and p_0 are the r.m.s. pressures corresponding to the intensities I_1 and I_0.

Loudness is the physiological counterpart of acoustical intensity. It is a function of the intensity but also varies with frequency and composition of the note being heard. The *Weber–Fechner Law* states that the sensation (loudness) is proportional to the logarithm of the stimulus (intensity).

Loudness level of a sound is judged by comparison in free air with a standard sinusoidal note whose frequency is 1000 Hz. The unit is the *phon*. If an average observer decides that a sound is equally loud as the standard 1000 Hz note of intensity n dB above the standard reference level corresponding to a r.m.s. pressure of 2×10^{-5} Pa (*i.e.* the threshold of hearing), then the sound is said to have an 'equivalent loudness' of n British Standard phons.

Reverberation in an enclosure is the persistence of sound due to multiple reflections from the walls, etc. of the enclosure.

Reverberation time is the time required, from the moment of cessation of a sound for the intensity to drop by 60 dB, *i.e.* to one millionth of its original value. Unit, *second*.

Absorption Coefficient of a surface is the ratio of the sound energy absorbed to the total sound energy incident on the surface. The ideal absorber is one from which no sound is reflected or scattered. For unit area of various substances, the coefficient is expressed in terms of equivalent area of open window (diffraction effects excluded). Unit, *ft² of open window* or *Sabine*. The coefficient varies with frequency.

Sabine's Relation: For an auditorium whose walls, etc. consist of areas $S_1 \, S_2 \ldots$ etc. of absorption coefficient $\alpha_1 \, \alpha_2 \ldots$ etc., the reverberation time t (in seconds) is given by $t = \dfrac{0 \cdot 05 V}{\Sigma \alpha \, S}$ (unit of length, ft) or $t = \dfrac{0 \cdot 16 V}{\Sigma \alpha \, S}$ (unit of length, metre) where V is the volume of the auditorium and $\Sigma \alpha \, S = \alpha_1 S_1 + \alpha_2 S_2 + \ldots$ etc.

8 The Periodic Table—giving atomic number and chemical symbol for each element

1 H																	2 He
3 Li	4 Be											5 B	6 C	7 N	8 O	9 F	10 Ne
11 Na	12 Mg											13 Al	14 Si	15 P	16 S	17 Cl	18 Ar
19 K	20 Ca	21 Sc	22 Ti	23 V	24 Cr	25 Mn	26 Fe	27 Co	28 Ni	29 Cu	30 Zn	31 Ga	32 Ge	33 As	34 Se	35 Br	36 Kr
37 Rb	38 Sr	39 Y	40 Zr	41 Nb	42 Mo	43 Tc	44 Ru	45 Rh	46 Pd	47 Ag	48 Cd	49 In	50 Sn	51 Sb	52 Te	53 I	54 Xe
55 Cs	56 Ba	57* La	72 Hf	73 Ta	74 W	75 Re	76 Os	77 Ir	78 Pt	79 Au	80 Hg	81 Tl	82 Pb	83 Bi	84 Po	85 At	86 Rn
87 Fr	88 Ra	89† Ac															

←————— TRANSITION ELEMENTS —————→

*LANTHANONS													
58 Ce	59 Pr	60 Nd	61 Pm	62 Sm	63 Eu	64 Gd	65 Tb	66 Dy	67 Ho	68 Er	69 Tm	70 Yb	71 Lu

†ACTINONS													
90 Th	91 Pa	92 U	93 Np	94 Pu	95 Am	96 Cm	97 Bk	98 Cf	99 Es	100 Fm	101 Md	102 No	103 Lr

9 The Arrangement of Electrons in Atoms

The table below gives the numbers of electrons in the various shells of the atom. It refers to neutral atoms in their lowest energy states. The usual notation is used for the shells. Thus, the number refers to the *principal* quantum number and the letter identifies the *orbital* or *azimuthal* quantum number. The letters: s, p, d, f, g, h, k etc. identify shells with orbital quantum numbers: 0, 1, 2, 3, 4, 5, 6 etc. Thus the 4s shell has principal quantum number, 4, and orbital quantum number, 0.

Electron Arrangement

Atomic Number	Element	K	L		M			N				O
		1s	2s	2p	3s	3p	3d	4s	4p	4d	4f	5s
1	H	1										
2	He	2										
3	Li	2	1									
4	Be	2	2									
5	B	2	2	1								
6	C	2	2	2								
7	N	2	2	3								
8	O	2	2	4								
9	F	2	2	5								
10	Ne	2	2	6								
11	Na	2	2	6	1							
12	Mg	2	2	6	2							
13	Al	2	2	6	2	1						
14	Si	2	2	6	2	2						
15	P	2	2	6	2	3						
16	S	2	2	6	2	4						
17	Cl	2	2	6	2	5						
18	Ar	2	2	6	2	6						
19	K	2	2	6	2	6		1				
20	Ca	2	2	6	2	6		2				
21	Sc	2	2	6	2	6	1	2				
22	Ti	2	2	6	2	6	2	2				
23	V	2	2	6	2	6	3	2				
24	Cr	2	2	6	2	6	5	1				
25	Mn	2	2	6	2	6	5	2				
26	Fe	2	2	6	2	6	6	2				
27	Co	2	2	6	2	6	7	2				
28	Ni	2	2	6	2	6	8	2				
29	Cu	2	2	6	2	6	10	1				
30	Zn	2	2	6	2	6	10	2				
31	Ga	2	2	6	2	6	10	2	1			
32	Ge	2	2	6	2	6	10	2	2			
33	As	2	2	6	2	6	10	2	3			
34	Se	2	2	6	2	6	10	2	4			
35	Br	2	2	6	2	6	10	2	5			
36	Kr	2	2	6	2	6	10	2	6			
37	Rb	2	2	6	2	6	10	2	6			1
38	Sr	2	2	6	2	6	10	2	6			2
39	Y	2	2	6	2	6	10	2	6	1		2
40	Zr	2	2	6	2	6	10	2	6	2		2
41	Nb	2	2	6	2	6	10	2	6	4		1
42	Mo	2	2	6	2	6	10	2	6	5		1
43	Tc	2	2	6	2	6	10	2	6	6		1
44	Ru	2	2	6	2	6	10	2	6	7		1
45	Rh	2	2	6	2	6	10	2	6	8		1
46	Pd	2	2	6	2	6	10	2	6	10		

Electron Arrangement

Atomic Number	Element	K	L		M			N				O					P					Q	
		1s	2s	2p	3s	3p	3d	4s	4p	4d	4f	5s	5p	5d	5f	5g	6s	6p	6d	6f	6g	7s	7p
47	Ag	2	2	6	2	6	10	2	6	10		1											
48	Cd	2	2	6	2	6	10	2	6	10		2											
49	In	2	2	6	2	6	10	2	6	10		2	1										
50	Sn	2	2	6	2	6	10	2	6	10		2	2										
51	Sb	2	2	6	2	6	10	2	6	10		2	3										
52	Te	2	2	6	2	6	10	2	6	10		2	4										
53	I	2	2	6	2	6	10	2	6	10		2	5										
54	Xe	2	2	6	2	6	10	2	6	10		2	6										
55	Cs	2	2	6	2	6	10	2	6	10		2	6				1						
56	Ba	2	2	6	2	6	10	2	6	10		2	6				2						
57	La	2	2	6	2	6	10	2	6	10		2	6	1			2						
58	Ce	2	2	6	2	6	10	2	6	10	2	2	6				2						
59	Pr	2	2	6	2	6	10	2	6	10	3	2	6				2						
60	Nd	2	2	6	2	6	10	2	6	10	4	2	6				2						
61	Pm	2	2	6	2	6	10	2	6	10	5	2	6				2						
62	Sm	2	2	6	2	6	10	2	6	10	6	2	6				2						
63	Eu	2	2	6	2	6	10	2	6	10	7	2	6				2						
64	Gd	2	2	6	2	6	10	2	6	10	7	2	6	1			2						
65	Tb	2	2	6	2	6	10	2	6	10	9	2	6				2						
66	Dy	2	2	6	2	6	10	2	6	10	10	2	6				2						
67	Ho	2	2	6	2	6	10	2	6	10	11	2	6				2						
68	Er	2	2	6	2	6	10	2	6	10	12	2	6				2						
69	Tm	2	2	6	2	6	10	2	6	10	13	2	6				2						
70	Yb	2	2	6	2	6	10	2	6	10	14	2	6				2						
71	Lu	2	2	6	2	6	10	2	6	10	14	2	6	1			2						
72	Hf	2	2	6	2	6	10	2	6	10	14	2	6	2			2						
73	Ta	2	2	6	2	6	10	2	6	10	14	2	6	3			2						
74	W	2	2	6	2	6	10	2	6	10	14	2	6	4			2						
75	Re	2	2	6	2	6	10	2	6	10	14	2	6	5			2						
76	Os	2	2	6	2	6	10	2	6	10	14	2	6	6			2						
77	Ir	2	2	6	2	6	10	2	6	10	14	2	6	9									
78	Pt	2	2	6	2	6	10	2	6	10	14	2	6	9			1						
79	Au	2	2	6	2	6	10	2	6	10	14	2	6	10			1						
80	Hg	2	2	6	2	6	10	2	6	10	14	2	6	10			2						
81	Tl	2	2	6	2	6	10	2	6	10	14	2	6	10			2	1					
82	Pb	2	2	6	2	6	10	2	6	10	14	2	6	10			2	2					
83	Bi	2	2	6	2	6	10	2	6	10	14	2	6	10			2	3					
84	Po	2	2	6	2	6	10	2	6	10	14	2	6	10			2	4					
85	At	2	2	6	2	6	10	2	6	10	14	2	6	10			2	5					
86	Rn	2	2	6	2	6	10	2	6	10	14	2	6	10			2	6					
87	Fr	2	2	6	2	6	10	2	6	10	14	2	6	10			2	6				1	
88	Ra	2	2	6	2	6	10	2	6	10	14	2	6	10			2	6				2	
89	Ac	2	2	6	2	6	10	2	6	10	14	2	6	10			2	6	1			2	
90	Th	2	2	6	2	6	10	2	6	10	14	2	6	10			2	6	2			2	
91	Pa	2	2	6	2	6	10	2	6	10	14	2	6	10	2		2	6	1			2	
92	U	2	2	6	2	6	10	2	6	10	14	2	6	10	3		2	6	1			2	

10 Properties of the Elements

The following table lists the elements with atomic number up to 92 alphabetically by name. Columns 1-4 and 13-16 are self-explanatory. Column 5 gives the crystal structures of the elements in their solid state. Where a change in structure occurs, the transition temperature is indicated (in K) under the crystal structures. The following abbreviations are used:

bcc = body-centred cubic
cubic (diam) = diamond structure
fcc = face-centred cubic
hcp = hexagonal close-packed

hex = hexagonal
mon = monoclinic
ortho = orthorhombic
tetra = tetragonal

Column 6 lists the atomic radii of the elements in pm (10^{-12} m). These radii are calculated as half the distance of closest approach of atomic centres in the crystalline state. Column 7 gives the principal oxidation numbers and column 8, the corresponding ionic radii. Columns 9 and 10 give the energies (eV) required to remove the first and second electrons from the atom—multiply by 96·49 to convert to kJ mol^{-1}. Column 11 gives the energy required to remove an electron from the negative ion formed by the atom with an extra electron. These 'electron affinities' are difficult to measure and there are few reliable results. Column 12 gives the electronegativities assigned to the elements by Pauling. These are numbers between 0 and 4 which may be used in determining the contribution of the ionic and covalent components of the bonds between different atoms.

Symbol	Name	Atomic Number Z	Atomic Weight M/g mol^{-1}	Crystal Structure	Atomic radius r_a/pm	Principal Oxidation Numbers	Ionic Radii r_i/pm	Ionization Energies E_i/eV		Electron Affinities E_e/eV	Electronegativities	Density ρ/kg m^{-3}	Melting Point T_m/K	Boiling Point T_b/K	Symbol
Ac	Actinium	89	227	fcc	188	3+	118	6·9	12·1	—	1·1	10 100	1 320	3 470	Ac
Al	Aluminium	13	26·98	fcc	142	3+	51	5·986	18·828	0·5	1·5	2 700	933·2	2 740	Al
Sb	Antimony	51	121·75	rhombic	145	3+ 5+	76	8·641	16·53	>2·0	1·9	6 700	903·7	1 650	Sb
Ar	Argon	18	39·95	fcc	174	0(1+)	154	15·759	27·629	-1·0	—	1·66	83·7	87·4 (sub)	Ar
As	Arsenic	33	74·92	rhombic	125	3+ 5+	58	9·81	18·633	—	2·0	5 730	1 090 (28 atm)	886 (sub)	As
At	Astatine	85	210	—	—	7+	62	9·5	—	—	2·2	—	520	623	At
Ba	Barium	56	137·34	bcc	217	2+	134	5·212	10·004	—	0·9	3 600	1 000	1 910	Ba
Be	Beryllium	4	9·01	hcp/cubic 1527	112	2+	35	9·322	18·211	0·30	1·5	1 800	1 550	3 243	Be
Bi	Bismuth	83	208·98	rhombic	155	3+ 5+	96	7·289	16·69	>0·7	1·9	9 800	544·4	1 830	Bi
B	Boron	5	10·81	ortho (?)	88	3+	23	8·298	25·154	0·33	2·0	2 500	2 600	2 820 (sub)	B
Br	Bromine	35	79·90	ortho	114	1- 5+	196 / 47	11·814	21·8	3·363	2·8	3 100 (298 K)	265·9	331·9	Br
Cd	Cadmium	48	112·40	hcp	148	2+	97	8·993	16·908	—	1·7	8 650	594·2	1 038	Cd
Cs	Caesium	55	132·90	bcc	262	1+	167	3·894	25·1	>0·19	0·7	1 870	301·6	960	Cs
Ca	Calcium	20	40·08	fcc/bcc 737	196	2+	99	6·113	11·871	—	1·0	1 540	1 120	1 760	Ca
C	Carbon	6	12·01	hex/cubic graph/diam	71/77 g/d	4+ 4-	16 / 260	11·260	24·383	1·25	2·5	2 300	>3 800	5 100	C

Ce	3 740	1 070	6 800	1·1	—	10·85	5·47	103, 92	3+, 4+	183	fcc/hex/fcc/bcc 95 263 998	140·12	58	Cerium · Ce
Cl	238·5	172·1	3·21 (273 K)	3·0	3·615	23·81	12·967	181, 34, 27	1−, 5+, 7+	91	tetra	35·45	17	Chlorine · Cl
Cr	2 755	2 160	7 200	1·6	0·98	16·50	6·766	63, 52	3+, 6+	125	bcc	52·00	24	Chromium · Cr
Co	3 170	1 765	8 900	1·8	0·9	17·06	7·86	72, 63	2+, 3+	125	hcp/fcc 690	58·93	27	Cobalt · Co
Cu	2 868	1 356	8 930	1·9	1·8	20·292	7·726	96, 72	1+, 2+	128	fcc	63·55	29	Copper · Cu
Dy	2 900	1 680	8 500	1·2	—	11·67	5·93	91	3+	175	rhombic/hcp 86	162·50	66	Dysprosium · Dy
Er	3 200	1 770	9 000	1·2	—	11·93	6·10	88	3+	173	hcp	167·26	68	Erbium · Er
Eu	1 712	1 100	5 200	1·1	—	11·25	5·67	95, 109	2+, 3+	198	bcc	151·96	63	Europium · Eu
F	85·01	53·5	1·7 (273 K)	4·0	3·448	34·97	17·422	133	1−	60	—	19·00	9	Fluorine · F
Fr	920	303	—	0·7	—	4·0	4·0	180	1+	—	—	223	87	Francium · Fr
Gd	3 000	1 585	7 900	1·2	—	12·1	6·14	94	3+	178	hcp/bcc 1537	157·25	64	Gadolinium · Gd
Ga	2 676	302·9	5 950	1·6	—	20·51	5·999	62	3+	121	fcc or ortho	69·72	31	Gallium · Ga
Ge	3 100	1 210·5	5 400	1·8	—	15·934	7·899	53	4+	122	cubic (diam)	72·59	32	Germanium · Ge
Au	3 239	1 336·1	19 300	2·4	2·1	20·5	9·225	137, 85	1+, 3+	144	fcc	196·97	79	Gold · Au
Hf	5 700	2 423	13 300	1·3	—	14·9	7·0	78	4+	158	hcp/bcc 2050	178·49	72	Hafnium · Hf
He	4·21	0·95 (26 atm)	0·166	—	−0·53	54·416	24·587	—	0	176	hcp/cubic	4·003	2	Helium · He
Ho	2 900	1 734	8 800	1·2	—	11·80	6·02	89	3+	176	hcp	164·93	67	Holmium · Ho
H	20·4	14·01	0·08987 (273 K)	2·1	0·76	—	13·598	154	1+	46	hcp/cubic	1·00797	1	Hydrogen · H
In	2 300	429·8	7 310	1·7	—	18·869	5·786	81	3+	162	bc tetra	114·82	49	Indium · In
I	457·4	386·6	4 940	2·5	3·070	19·131	10·451	216	1−	135	ortho	126·90	53	Iodine · I
Ir	4 800	2 716	22 420	2·2	—	16·18	9·1	68	4+	135	fcc	192·2	77	Iridium · Ir
Fe	3 300	1 808	7 870	1·8	0·6	—	7·87	74, 64	2+, 3+	123	bcc/fcc/bcc 1180 1670	55·85	26	Iron · Fe
Kr	120·8	116·5	3·49	—	—	24·359	13·999	—	0	201	fcc	83·80	36	Krypton · Kr
La	3 742	1190	6 150	1·1	—	11·06	5·577	102	3+	187	hcp/fcc/bcc 583 1137	138·91	57	Lanthanum · La
Pb	2 017	600·4	11 340	1·8	—	15·032	7·416	120, 84	2+, 4+	174	fcc	207·19	82	Lead · Pb
Li	1 590	452	534	1·0	0·82	75·638	5·392	68	1+	152	hcp/fcc/bcc 74 140	6·94	3	Lithium · Li

58

Name	Symbol	Atomic Number Z	Atomic Weight M/g mol^{-1}	Crystal Structure	Atomic radius r_a/pm	Principal Oxidation Numbers	Ionic Radii r_i/pm	Ionization Energies E_i/eV	Electron Affinities E_e/eV	Electronegativities	Density ρ/kg m^{-3}	Melting Point T_m/K	Boiling Point T_b/K	Symbol
Lutetium	Lu	71	174·97	hcp	173	3+	85	5·426 13·9 14·7	—	1·2	9 800	1 925	3 600	Lu
Magnesium	Mg	12	24·31	hcp	160	1+ 2+	82 66	7·646 15·035	-0·32	1·2	1 741	924	1 380	Mg
Manganese	Mn	25	54·94	cubic	112	2+ 3+	80 66	7·435 15·640	—	1·5	7 440	1 517	2 370	Mn
Mercury	Hg	80	200·59	rhombic	156	1+ 2+	127 110	10·437 18·756	1·54	1·9	13 590 (273 K)	234·3	629·7	Hg
Molybdenum	Mo	42	95·94	bcc	136	2+ 4+ 6+	70 62 100	7·099 16·15	1·0	1·8	10 200	2 880	5 830	Mo
Neodymium	Nd	60	144·24	hcp/bcc 1135	181	3+	69	5·49 10·72	—	1·1	6 960	1 297	3 300	Nd
Neon	Ne	10	20·18	fcc	160	0	—	21·564 40·962	-0·57	—	0·839	24·5	27·2	Ne
Nickel	Ni	28	58·71	fcc	124	2+ 3+	69	7·635 18·168	1·3	1·8	8 900	1 726	3 005	Ni
Niobium	Nb	41	92·91	bcc	143	5+	69 16	6·88 14·534	—	1·6	8·570	2 741	5 200	Nb
Nitrogen	N	7	14·01	cubic/hcp 35·4	71	3+ 5+	13 69	14·32 29·601	0·05	3·0	1·165	63·3	77·3	N
Osmium	Os	76	190·2	hcp	135	4+	69	8·7 13·618	—	2·2	22 480	3 300	4 900	Os
Oxygen	O	8	16·00	rhombic	60	2-	132	17·0 35·116	1·471	3·5	1·33	54·7	90·2	O
Palladium	Pd	46	106·4	fcc	137	2+ 4+	80 65	8·34 19·43	—	2·2	12 000	1 825	3 200	Pd
Phosphorus	P	15	30·97	cubic	—	3+ 5+	44 35	10·486 19·725	0·8	2·1	2 200 (r) 1 800 (y)	317·2	552	P
Platinum	Pt	78	195·09	fcc	138	2+ 4+	80 65	9·0 18·563	—	2·2	21 450	2 042	4 100	Pt
Polonium	Po	84	209	monoclinic	168	2+ 6+	67	8·42 19·4	—	2·0	9 400	527	1 235	Po
Potassium	K	19	39·10	bcc	231	1+	133	4·341 31·625	0·82	0·8	860	336·8	1 047	K
Praseodymium	Pr	59	140·91	hcp/bcc 1065	182	3+	101	5·42 10·55	—	1·1	6 800	1 208	3 400	Pr
Promethium	Pm	61	145	—	—	3+	98	5·55 10·90	—	1·1	—	1 308	3 000	Pm
Protoactinium	Pa	91	231	tetra	160	3+ 4+	113 98	—	—	1·5	15 400	1 500	4 300	Pa

Symbol	Name														Symbol
Ra	Radium	88	226	—	—	2+	143	5·279	10·147	—	0·9	5 000	970	1 410	Ra
Rn	Radon	86	222	—	—	0	—	10·748	—	—	—	9·73 (273 K)	202	211·3	Rn
Re	Rhenium	75	186·2	hcp	137	4+	72	7·88	16·6	—	1·9	20 500	3 450	5 900	Re
Rh	Rhodium	45	102·91	fcc	134	3+	68	7·46	18·08	0·15	2·2	12 440	2 230	4 000	Rh
Rb	Rubidium	37	85·47	bcc	246	1+	147	4·177	27·28	—	0·8	1 530	312·0	961	Rb
Ru	Ruthenium	44	101·07	hcp	133	4+	67	7·374	16·76	0·4	2·2	12 400	2 520	4 200	Ru
Sm	Samarium	62	150·35	Rhomb/bcc 1190	179	3+	96	5·63	11·07	—	1·1	7 500	1 345	2 200	Sm
Sc	Scandium	21	44·96	hcp/fcc 1223	160	3+	73	6·54	12·80	—	1·3	3 000	1 812	3 000	Sc
Se	Selenium	34	78·96	hcp	116	2−	191	9·752	21·19	3·7	2·4	4 810	490	958	Se
Si	Silicon	14	28·09	cubic	118	4+	42	8·151	16·345	1·5	1·8	2 300	1 680	2 628	Si
Ag	Silver	47	107·87	fcc/hcp 5	144	1+	126	7·576	21·49	2·5	1·9	10 500	1 234	2 485	Ag
Na	Sodium	11	22·99	bcc	185	1+	95	5·139	47·286	0·84	0·9	970	371	1 165	Na
Sr	Strontium	38	87·62	fcc/hcp/bcc 506 813	215	2+	113	5·695	11·030	—	1·0	2 600	1042	1 657	Sr
S	Sulphur	16	32·06	fc ortho	106	2−/4+/6+	184	10·360	23·33	2·07	2·5	2 070	386	717·7	S
Ta	Tantalum	73	180·95	bcc	143	5+	68	7·89	16·2	—	1·5	16 600	3 269	5 698	Ta
Tc	Technetium	43	98·91	hcp	135	7+	98	7·28	15·26	—	1·9	11 400	2 500	4 900	Tc
Te	Tellurium	52	127·60	hcp	143	2−/4+/6+	211	9·009	18·6	3·6	2·1	6 240	722·6	1 260	Te
Tb	Terbium	65	158·92	hcp/rhomb 1590	177	3+	92	5·85	11·52	—	1·2	8 300	1 629	3 100	Tb
Tl	Thallium	81	204·37	hcp/fcc 503	171	1+/3+	147	6·108	20·428	—	1·8	11 860	576·6	1 730	Tl
Th	Thorium	90	232·04	fcc/bcc 1673	180	4+	102	6·95	11·5	—	1·3	11 500	2 000	4 500	Th
Tm	Thulium	69	168·93	hcp/bcc 1158	174	3+	87	6·18	12·05	—	1·2	9 300	1 818	2 000	Tm
Sn	Tin	50	118·69	cub(diam)/bcc	140	2+/4+	93	7·344	14·632	—	1·8	7 300	505·1	2 540	Sn
Ti	Titanium	22	47·90	hcp/bcc 1158	146	3+/4+	68	6·82	13·58	0·39	1·5	4 540	1 948	3 530	Ti
W	Tungsten	74	183·85	bcc	137	6+	62	7·98	17·7	0·5	1·7	19 320	3 650	6 200	W
U	Uranium	92	238·03	rhomb/tetr 941	138	4+/6+	97	6·08	—	0·94	1·7	19 050	1 405·4	4 091	U
V	Vanadium	23	50·94	bcc	131	3+/5+	59	6·74	14·65	—	1·6	6 100	2 160	3 300	V
Xe	Xenon	54	131·30	fcc	221	0	—	12·130	21·21	—	—	5·50	161·2	166·0	Xe
Yb	Ytterbium	70	173·04	fcc/bcc 1071	193	3+	86	6·254	12·17	—	1·2	7 000	1 097	1 700	Yb
Y	Yttrium	39	88·91	hcp/bcc 1763	181	3+	89	6·38	12·24	—	1·2	4 600	1 768	3 200	Y
Zn	Zinc	30	65·37	hcp	133	2+	74	9·394	17·964	1·6	1·6	7 140	692·6	1 180	Zn
Zr	Zirconium	40	91·22	hcp/bcc 1100	160	4+	79	6·84	13·13	1·4	1·4	6 500	2 125	3 851	Zr

11 Properties of Metallic Solids (at 293 K)

Values quoted for Tensile Strength and Yield Stress are in units of 10^6 N m^{-2}($=$ MPa). Values of Young's Modulus are in units of 10^9 N m^{-2}($=$ GPa). These values are typical observations and are approximate only. The elastic properties vary somewhat between specimens depending on the manufacturing process and the previous history of the sample. The Shear Modulus (G) and Bulk Modulus (K) can be calculated from the relations: $G = \frac{1}{2}E/(1 + \nu)$ and $K = \frac{1}{3}E/(1 - 2\nu)$, where E is Young's Modulus and ν is Poisson's Ratio.

Name	Density ρ/kg m^{-3}	Melting Point T_m/K	Specific Latent Heat of Fusion l/J kg^{-1} $\times 10^4$	Specific Heat Capacity c_p/J kg^{-1} K^{-1}	Linear Expansivity α/K^{-1} $\times 10^{-6}$	Thermal Conductivity λ/W m^{-1} K^{-1}	Electrical Resistivity ρ/Ω m $\times 10^{-8}$	Temperature Coefficient of Resistance $(1/\rho_0)(d\rho/dT)$/K^{-1} $\times 10^{-4}$	Tensile Strength σ_T/MPa	Yield Strength σ_Y/MPa	Elongation e/%	Young's Modulus E/GPa	Poisson's Ratio ν	
1 Aluminium	2 710	932	38	913	23	201	2·65	40	80	50	43	71	0·34	1
2 Aluminium, strong alloy	2 800	800	39	880	23	180	5	16	600	550	10			2
3 Antimony	6 680	904	16	205	10	18	40	~50				78		3
4 Bismuth	9 800	544	5	126	13	8	115	45				32	0·33	4
5 Brass (70Cu/30Zn)	8 500	1300		370	18	110	~8	~15	550	450	8	100	0·35	5
6 Bronze (90Cu/10Sn)	8 800	1300		360	17	180	30		260	140	10			6
7 Cobalt	8 900	1765	25	420	12	69	6	66	~500					7
8 Constantan	8 880	1360		420	17	23	47	±0·4				170	0·33	8
9 Copper	8 930	1356	21	385	17	385	1·7	39	150	75	45	117	0·35	9
10 German silver (60Cu/25Zn/15Ni)	8 700	1300		400	18	29	33	4	450			130	0·33	10
11 Gold	19 300	1340	7	132	14	296	2·4	34	120		40	71	0·44	11
12 Invar (64Fe/36Ni)	8 000	1800		503	0·9	16	81	20	480	280	40	145	0·26	12
13 Iron, pure	7 870	1810	27	106	12	80	10	65	300	165	45	206	0·29	13
14 Iron, cast grey	7 150	1500	10	500	11	75	10		100			110	0·27	14
15 ,, ,, white	7 700	1420	14		11	75	10		230		~0			15
16 ,, wrought	7 850	1810	14	480	12	60	14	60	~370	150	45	197	0·28	16
17 Lead	11 340	600	2·6	126	29	35	21	43	15	12	50	18	0·44	17
18 Magnesium	1 740	924	38	246	25	150	4	43	190	95	5	44	0·29	18
19 Manganin	8 500		41	400	18	22	45	±0·1				120	0·33	19
20 Monel (70Ni/30Cu)	8 800	1600			14	210	42	20	520	240	40			20
21 Nickel	8 900	1726	31	460	13	59	59	60	300	60	30	207	0·36	21
22 Nickel, strong alloy	8 500	1320		380					1300	1200	10	110	0·38	22
23 Phosphor bronze					17		7	60	560	420		120	0·38	23
24 Platinum	21 450	2042	11	136	9	69	11	38	350			150	0·38	24
25 Silver	10 500	1230	10	235	19	419	1·6	40	150	180	45	70	0·37	25
26 Sodium	970	371	12	1240	71	134	4·5	44						26
27 Solder, soft (50Pb/50Sn)	9 000	490	190	176					45		50			27
28 Stainless Steel (18Cr/8Ni)	7 930	1800		510	16	150	96	6	600	230	60			28
29 Steel, mild	7 860	1700		420	15	63	15	50	460	300	35	210	0·29	29
30 Steel, piano wire	7 800	1700				50			3000			210	0·29	30
31 Tin	7 300	505	6·0	226	23	65	11	50	30			40	0·36	31
32 Titanium	4 540	1950		523	9	23	53	38	620	480	20		0·36	32
33 Zinc	7 140	693	10	385	31	111	5·9	40	150		50	110	0·25	33

The following table lists materials which do not readily conduct electricity. In many cases the physical constants cannot be specified accurately as the values observed depend so much on the manufacture and life history of the specimen. The values given are to be taken as representative only.

Name	Density ρ/kg m⁻³	Melting Point T_m/K	Specific Heat Capacity c_p/J kg⁻¹ K⁻¹	Linear Expansivity α/K⁻¹ ×10⁻⁶	Thermal Conductivity λ/W m⁻¹ K⁻¹	Tensile Strength σ_T/MPa	Elongation e/%	Young's Modulus E/GPa	
1 Alumina, ceramic	3 800	2300	800	9	29	~150		345	1
2 Bone	1 850					140		28	2
3 Brick, building	2 300			9	0·6	~5			3
4 fireclay	2 100			4·5	0·8				4
5 paving	2 500			4·0					5
6 silica	1 750				0·8				6
7 Carbon, graphite	2 300	3800	710	7·9	5·0			207	7
8 diamond	3 300		525	~0	900			1200	8
9 Concrete	2 400		3350	12	0·1	~4		14	9
10 Cork	240		2050		0·05				10
11 Cotton	1 500		1400			400			11
12 Epoxy resin	1 120		1400	39		50	2–6	4·5	12
13 Fluon (PTFE)	2 200		1050	55	0·25	22	50–75	0·34	13
14 Glass (crown)	2 600	1400	670	9	1·0	~100		71	14
15 (flint)	4 200	1500	500	8	0·8			80	15
16 Glass wool	50	1400	670		0·04				16
17 Ice	920	273	2100	51	2·0				17
18 Kapok	50				0·03				18
19 Magnesium oxide	3 600	3200	960	12				207	19
20 Marble	2 600		880	10	2·9				20
21 Melamine formaldehyde	1 500		1700	40	0·3	70		9	21
22 Naphthalene	1 150	350	1310	107	0·4				22
23 Nylon	1 150	470	1700	100	0·25	70	60–300		23
24 Paraffin wax	900	330	2900	110	0·25				24
25 Perspex	1 190	350	1500	85	0·2	50	2–7	3	25
26 Phenol formaldehyde	1 300		1700	40	0·2	50	0·4–0·8	6·9	26
27 Polyethylene (low den)	920	410	2300	250		13	400–800	0·18	27
28 (high den)	955	410	2300	250		26	100–300	0·43	28
29 Polypropylene	900	450	2100	62		35	>220	1·2	29
30 Polystyrene	1 050	510	1300	70	0·08	50	1–3	3·1	30
31 Polyvinylchloride (non-rigid)	1 250	485	1800	150		15	200–400	0·01	31
32 (rigid)	1 700	485	1000	55		60	5–25	2·8	32
33 Polyvinylidine chloride		470		190		30	160–240		33
34 Quartz fibre	2 660	2020	788	0·4	9·2			73	34
35 Rubber (polyisoprene)	910	300	1600	220	0·15	17	480–510	0·02	35
36 Silicon carbide	3 170			4·5					36
37 Sulphur	2 070	386	730	64	0·26				37
38 Titanium carbide	4 500			7	28			345	38
39 Wood, oak (with grain)	650				0·15			12	39
40 ,, Spruce (with grain)	600							14	40
41 ,, Spruce (across grain)								0·5	41

13 Properties of Liquids (at 293 K)

	Name	Density ρ/kg m^{-3}	Melting Point T_m/K	Boiling Point T_b/K	Specific Latent Heat of Vaporization l/J kg^{-1} $\times10^4$	Specific Heat Capacity c_p/J kg^{-1} K^{-1}	Cubic Expansivity γ/K^{-1} $\times10^{-4}$	Thermal Conductivity λ/W m^{-1} K^{-1}	Surface Tension σ/N m^{-1} $\times10^{-3}$	Viscosity η/N s m^{-2} $\times10^{-3}$	Refractive Index n	Bulk Modulus of Rigidity K/GPa
1	Acetic acid ($C_2H_4O_2$)	1049	290	391	39	1960	10·7	0·180	27·6	1·219	1·3718	2·49
2	Acetone (C_3H_6O)	780	178	330	52	2210	14·3	0·161	23·7	0·324	1·3620 (288 K)	~0·8
3	Benzene (C_6H_6)	879	279	353	40	1700	12·2	0·140	28·9	0·647	1·5011	1·10
4	Bromine (Br)	3100	266	332	18·3	460	11·3		41·5	0·993	1·66	1·58
5	Carbon disulphide (CS_2)	1293	162	319	36	1000	11·9	0·144	32·3	0·375	1·6276	1·16
6	Carbon tetrachloride (CCl_4)	1632	250	350	19	840	12·2	0·103	26·8	0·972	1·4607	1·12
7	Chloroform ($CHCl_3$)	1490	210	334	25	960	12·7	0·121	27·1	0·569	1·4467	1·1
8	Ether, diethyl ($C_4H_{10}O$)	714	157	308	35	2300	16·3	0·127	17	0·242	1·3538	0·69
9	Ethyl alcohol (C_2H_6O)	789	156	352	85	2500	10·8	0·177	22·3	1·197	1·3610	1·32
10	Glycerol ($C_3H_8O_3$)	1262	293	563	83	2400	4·7	0·270	63	1495	1·4730	4·03
11	Mercury (Hg)	13546	234	630	29	140	1·82	7·96	472	1·552	1·73	26·2
12	Methyl alcohol (CH_4O)	791	179	337	112	2500	11·9	0·201	22·6	0·594	1·3276	0·97
13	Nitrobenzene ($C_6H_5NO_2$)	1175	279	484	33	1400	8·6	0·160	43·9	2·03	1·5530	2·2
14	Olive oil	920		570		1970	7·0	0·170	32	84	1·48	1·60
15	Paraffin oil	800			53	2130	900	0·150	26	~1000	1·43	1·62
16	Phenol (C_6H_6O)	1073	314	455		2350	7·9		40·9	12·74	1·5425 (313 K)	
17	Toluene (C_7H_8)	867	178	384	35	1670	10·7	0·134	28·4	0·585	1·4969	1·09
18	Turpentine	870	263	429	29	1760	9·7	0·136	27	1·49	1·48	1·28
19	Water (H_2O)	998	273	373	226	4190	2·1	0·591	72·7	1·000	1·333	2·05
20	Water, sea	1025	264	~377		3900					1·343	

14 Properties of Gases at S.T.P.

Substance	Density ρ/kg m⁻³	Boiling Point T_b/K	Specific Latent Heat of Vaporization l/J kg⁻¹ ×10⁴	Specific Heat Capacity c_p/J kg⁻¹ K⁻¹	Ratio of Specific Heats $\gamma = (c_p/c_v)$	Thermal Conductivity λ/W m⁻¹ K⁻¹ ×10⁻⁴	Viscosity η/N s m⁻² ×10⁻⁶	Refractivity $(n-1)$ ×10⁻⁶	Critical Temperature T_c/K	Critical Pressure P_c/MPa	Critical Volume V_c/m³ mol⁻¹ ×10⁻⁶	
1 Acetylene (C_2H_2)	1·173	189	137·1	1590	1·26	184	9·35	606	309	6·14	113*	1
2 Air	1·293	83	21·4	993	1·402	241	18·325 (300 K)	292	132	3·77		2
3 Ammonia (NH_3)	0·771	240	137·1	2190	1·310	218	9·18	376	405	11·3	72·5	3
4 Argon (Ar)	1·784	87	15·8	524	1·667	162	21	281	151	4·86	75·2	4
5 Carbon dioxide (CO_2)	1·977	195	36·4	834	1·304	145	14	451	304	7·38	94·0	5
6 Carbon monoxide (CO)	1·250	81	21·1	1050	1·404	232	16·6	338	134	3·50	93·1	6
7 Chlorine (Cl_2)	3·214	238	28·1	478	1·36	72	12·9	773	417	7·71	124	7
8 Cyanogen (C_2N_2)	2·337	252	43·2	1720	1·26		9·28	835	401	6·0		8
9 Ethylene (C_2H_4)	1·260	170	48·4	1500	1·26	164	9·7	696	283	5·12	127·4	9
10 Helium (He)	0·179	4·25	2·5	5240	1·66	1415	18·6	36	5·3	0·23	58	10
11 Hydrogen (H_2)	0·090	20-35	45·3	14300	1·41	1684	8·35	132	33·3	1·294	65·5	11
12 Hydrogen chloride (HCl)	1·640	189	41·4	796	1·40		13·8	447	325	8·26	87	12
13 Hydrogen sulphide (H_2S)	1·538	211	55·3	1020	1·32	120	11·7	634	374	9·01	97·9	13
14 Methane (CH_4)	0·717	109	51·1	2200	1·313	302	10·3	444	191	4·62	98·7	14
15 Nitric oxide (NO)	1·340	121	46·2	972	1·394	238	17·8	297	179	6·5		15
16 Nitrogen (N_2)	1·250	77	20·9	1040	1·404	243	16·7	297	126	3·39	90·1	16
17 Nitrous oxide (N_2O)	1·978	183	37·6	892	1·303	151	13·5	516	310	7·24	96·7	17
18 Oxygen (O_2)	1·429	90	24·3	913	1·40	244	19·2	272	154	5·08	78	18
19 Sulphur dioxide (SO_2)	2·927	263	40·3	645		77	11·7	686	430	7·88	122	19
20 Water vapour (273 K) (H_2O)	0·800		226·1	2020 (373 K)	1·26	158	8·7	254	647	22·12	56·8	20

*The critical volume is here defined as the volume of one mole of the gas at its critical temperature and pressure.

THE MOHS SCALE OF HARDNESS

Substance	Hardness	Substance	Hardness	Substance	Hardness
Talc	1	Felspar	6	Fused zirconia	11
Gypsum	2	Vitreous silica	7	Fused alumina	12
Calcite	3	Quartz	8	Silicon carbide	13
Fluorite	4	Topaz	9	Boron carbide	14
Apatite	5	Garnet	10	Diamond	15

APPROXIMATE HARDNESS OF SOME COMMON MATERIALS

Substance	Hardness	Substance	Hardness	Substance	Hardness
Agate	6–7	Calcium	1·5	Glass	4·5–6·5
Aluminium	2–3	Carborundum	9–10	Marble	3–4
Amber	2–2·5	Chromium	9	Penknife blade	6·5
Asbestos	5	Copper	2·5–3	Silver	2·5–2·7
Brass	3–4	Finger nail	2·5	Steel (mild)	4–5

VISCOSITIES OF LIQUIDS AND THEIR TEMPERATURE DEPENDENCE, $\eta/\mathrm{N\,s\,m^{-2}}$

Substance	0°C	10°C	20°C	30°C	40°C	50°C
Water	0·001787	0·001304	0·001002	0·00080	0·000653	0·000547
Aniline	0·0102	0·0065	0·0044	0·00316	0·00237	0·00185
Benzene	0·000912	0·000758	0·000652	0·000564	0·000503	0·000442
Ethanol	0·00177	0·00147	0·0012	0·00100	0·000834	0·00070
Glycerol (propane-1,2,3-triol)	10·59	3·44	1·34	0·629	0·289	0·141
Rape oil	2·53	0·385	0·163	0·096	—	—

16 Electrical and Magnetic Data

IMPERIAL STANDARD WIRE GAUGE (SWG) AND WIRE RESISTANCES

Gauge Number	Diameter mm	Sectional Area mm²	Copper Ohm per metre	Eureka Ohm per metre	German Silver Ohm per metre	Manganin Ohm per metre	Nichrome Ohm per metre	Gauge Number
10	3·251	8·3019	0·00208	0·0590	0·0273	0·0500	0·130	10
12	2·642	5·4805	0·00315	0·0894	0·0413	0·0757	0·197	12
14	2·032	3·2429	0·00532	0·151	0·0698	0·128	0·333	14
16	1·626	2·0755	0·00831	0·236	0·109	0·200	0·520	16
18	1·219	1·1675	0·0148	0·420	0·194	0·355	0·925	18
20	0·9144	0·6567	0·0263	0·746	0·345	0·632	1·64	20
22	0·7112	0·3973	0·0434	1·23	0·570	1·04	2·72	22
24	0·5588	0·2453	0·0703	2·00	0·923	1·69	4·40	24
26	0·4572	0·16417	0·105	2·98	1·38	2·53	6·58	26
28	0·3759	0·11099	0·155	4·41	2·04	3·74	9·73	28
30	0·3150	0·07791	0·221	6·29	2·91	5·33	13·9	30
32	0·2743	0·05910	0·292	8·29	3·83	7·02	18·3	32
34	0·2337	0·04289	0·402	11·4	5·28	9·68	25·2	34
36	0·1930	0·02927	0·589	16·7	7·74	14·2	36·9	36
38	0·1524	0·018241	0·946	26·9	12·4	22·8	59·2	38
40	0·1219	0·011675	1·48	42·0	19·4	35·5	92·5	40
42	0·1016	0·008107	2·13	60·4	27·9	51·2	133	42
44	0·0813	0·005189	3·32	94·4	43·7	80·0	208	44
46	0·0610	0·002919	5·91	168	77·6	142	370	46
48	0·0406	0·0012972	13·3	378	175	320	833	48
50	0·0254	0·0005067	34·0	967	447	819	2130	50

PREFERRED METRIC SIZES

These are given in three series: R10, R20 and R40. Where possible, the R10 series is to be used, the intermediate sizes occurring in R20 and R40 being reserved for special purposes. See British Standard BS3737 (1964). The table gives the diameters of wires in the R40 series expressed in mm; alternate values form the second choice, R20 series, alternate values of which give the first choice, R10 series.

R10, R20	0·020	0·040	0·080	0·160	0·315	0·63	1·25	2·5	5·0	10·0	20·0
	0·021	0·042	0·085	0·170	0·335	0·67	1·32	2·65	5·3	10·6	21·2
R20	0·022	0·045	0·090	0·180	0·355	0·71	1·40	2·8	5·6	11·2	22·4
	0·024	0·048	0·095	0·190	0·375	0·75	1·50	3·0	6·0	11·8	23·6
R10, R20	0·025	0·050	0·100	0·200	0·40	0·80	1·60	3·15	6·3	12·5	25·0
	0·026	0·053	0·106	0·212	0·425	0·85	1·70	3·35	6·7	13·2	
R20	0·028	0·056	0·112	0·224	0·45	0·90	1·80	3·55	7·1	14·0	
	0·030	0·060	0·118	0·236	0·475	0·95	1·90	3·75	7·5	15·0	
R10, R20	0·032	0·063	0·125	0·250	0·50	1·00	2·00	4·0	8·0	16·0	
	0·034	0·067	0·132	0·265	0·53	1·06	2·12	4·25	8·5	17·0	
R20	0·036	0·071	0·140	0·28	0·56	1·12	2·24	4·5	9·0	18·0	
	0·038	0·075	0·150	0·30	0·60	1·18	2·36	4·75	9·5	19·0	

METRIC WIRE SIZES AND WIRE RESISTANCES

Wire Dia., mm	Sectional Area mm²	Copper Ohm per metre	Eureka Ohm per metre	German Silver Ohm per metre	Manganin Ohm per metre	Nichrome Ohm per metre	Wire Dia., mm
0·020	0·0003142	54·9	1560	721	1320	3440	0·020
0·025	0·0004909	35·1	998	461	845	2200	0·025
0·032	0·0008042	21·4	609	282	516	1340	0·032
0·040	0·001257	13·7	390	180	330	859	0·040
0·050	0·001963	8·79	250	115	211	550	0·050
0·063	0·003117	5·53	157	72·7	133	346	0·063
0·080	0·005027	3·43	97·5	45·1	82·6	215	0·080
0·100	0·007854	2·20	62·4	28·8	52·8	138	0·100
0·125	0·01227	1·41	39·9	18·5	33·8	88·7	0·125
0·160	0·02011	0·858	24·4	11·3	20·6	53·7	0·160
0·200	0·03142	0·549	15·6	7·21	13·2	34·4	0·200
0·250	0·04909	0·351	10·0	4·61	8·45	22·0	0·250
0·315	0·07793	0·221	6·29	2·91	5·33	13·9	0·315
0·400	0·1257	0·137	3·90	1·80	3·30	8·59	0·400
0·500	0·1963	0·0879	2·50	1·15	2·11	5·50	0·500

EMF OF STANDARD CELLS

Weston (Cadmium) cell (20°C)	= 1·0186 volts (absolute)
	= 1·0183 volts (international)
Clark cell (15°C)	= 1·4333 volts (absolute)
	= 1·4328 volts (international)

Temperature dependence
Weston cell
$E_t = 1·0186 - 0·0000406(t-20) - 9·5 \times 10^{-7}(t-20)^2$ absolute volts
Clark cell
$E_t = 1·4333 - 0·00119(t-15) - 7 \times 10^{-6}(t-15)^2$ absolute volts

APPROXIMATE EMFS OF CELLS

Bichromate	2	volts	Accumulator 2·0 volts (Ranges 1·85–2·2 volts)	
Bunsen	1·9	„	Dry cell	1·5 volts
Daniell	1·08	„	Nickel-Cadmium	1·3 „
Grove	1·8	„	Nickel-Iron .	1·4 „
Leclanché	1·46	„	Zinc-Silver oxide	1·8 „

RELATIVE PERMITTIVITIES (ε_r) OF VARIOUS SUBSTANCES AT ROOM TEMPERATURE (293 K)

Solid	ϵ_r	Liquid	ϵ_r	Gas	ϵ_r
Amber	2·8	Acetone	21·3	Air	1·000536
Ebonite	2·7–2·9	Benzene	2·28	Argon	1·000545
Glass	5–10	Carbon tetrachloride	2·17	Carbon dioxide	1·000986
Ice (268 K)	75	Castor oil	4·5	Carbon monoxide	1·00070
Marble	8·5	Ether	4·34	Deuterium	1·000270
Mica	5·7–6·7	Ethyl alcohol	25·7	Helium	1·00007
Paraffin wax	2–2·3	Glycerine	43	Hydrogen	1·00027
Perspex	3·5	Medicinal paraffin	2·2	Neon	1·000127
Polystyrene	2·55	Nitrobenzene	35·7	Nitrogen	1·000580
P.V.C.	4·5	Pentane	1·83	Oxygen	1·00053
Shellac	3–3·7	Silicon oil	2·2	Sulphur dioxide	1·00082
Sulphur	3·6–4·3	Turpentine	2·23	Water vapour	1·00060
Teflon	2·1	Water	80·37	(393 K)	

Values given in the table above refer to low frequencies, gases at 1 atmosphere pressure.

TEMPERATURE—EMF DATA FOR THERMOCOUPLES

The table gives the emf in millivolts for 'hot junction' temperatures from 0°–100°C. The 'cold junction' is maintained at 0°C.

Thermocouple	0°	10°	20°	30°	40°	50°	60°	70°	80°	90°	100°
Platinum—Platinum (90%), Rhodium (10%)	0	0·06	0·11	0·17	0·23	0·30	0·36	0·43	0·50	0·57	0·64
Copper—Constantan	0	0·39	0·79	1·19	1·61	2·03	2·47	2·91	3·36	3·81	4·28
Iron—Constantan	0	0·52	1·05	1·58	2·12	2·66	3·20	3·75	4·30	4·85	5·40

The mass susceptibility is given by the expression, $\chi_m = (\mu_r - 1)/\rho$; where μ_r is the relative permeability, and ρ the density of the specimen.

	χ_m/m^3		χ_m/m^3		χ_m/m^3
	$\times 10^{-8}$		$\times 10^{-8}$		$\times 10^{-8}$
Aluminium	$+0.82$	Glass	-1.3	Oxygen	$+133.6$
Araldite	-0.63	Helium	-0.59	Perspex	-0.5
Carbon (graphite)	-4.4	Hydrogen	-2.49	Polyethylene	$+0.2$
Copper	-0.108	Lead chloride	-0.40	P.V.C.	-0.75
Copper sulphate	$+7.7$	Manganese chloride	$+134$	Sodium chloride	-0.63
Ebonite	$+0.75$	Manganese dioxide	$+48.3$	Sulphur	-0.62
Iron ammonium alum	$+38.2$	Manganese sulphate	$+111$	Sulphuric acid	-0.50
Ferric hydroxide	$+197$	Mercury	-0.21	Water	-0.90
Ferrous sulphate	$+52.2$	Nitrogen	-0.54		

MAGNETIC PROPERTIES OF SOME 'SOFT' MAGNETIC MATERIALS

Alloy	Maximum Relative permeability $\mu_{r\,max}$	Coercive force $H_c/A\,m^{-2}$	Energy loss per cycle $E/J\,m^{-3}$	Resistivity $\rho/(ohm\,m)$	Saturation Induction B_m/T	Remarks
Iron, pure (total impurities $<0.005\%$)	200 000	4.0	30	10	2.15	commercially impracticable
Mild steel	2 000	143	-500	10		
Silicon iron (1.25% Si)	6 100	67.6	220			isotropic
Silicon iron (4.25% Si)	9 000	23.9	70	60	2.0	isotropic
Silicon iron (3% Si)	40 000	12	30	47	2.0	anisotropic, (110) ·100
Silicon iron (3.8% Si)	1 400 000		<3			single crystal
Silicon iron (6.3% Si)	500 000		4.5			polycrystalline, magnetically annealed: brittle
78 Permalloy (Fe21.5%Ni78.5%)	100 000	4.0			1.08	
Supermalloy (Fe16%Ni79% Mo5%)	1 000 000	0.16		60	0.79	
Ferroxcube 3 (Mn–Zn ferrite)	1 500	0.8		10^5	0.25	

PROPERTIES OF SOME COMMERCIAL PERMANENT MAGNET MATERIALS

Alloy	Compositon					Remanance B_r/T	Coercivity $_BH_c/A\,m^{-1}$	Maximum $B \times H$ $(BH)_{max}/J\,m^{-3}$	Comments
	Al	Ni	Co	Cu	Nb				
Alnico IV H	12	26	8	2		0.6	63 000	13×10^3	isotropic
Ticonal C	8	13.5	24	3	0.6	1.26	52 000	430	isotropic
Columax	8	13.5	24	3	0.5	1.35	64 000	64	columnar
Pt-Co alloy			23			0.45	210 000	300	ductile
Barium Ferrite (BaO. 6Fe₂O₃)						0.2	135 000	7 550	isotropic
Co₅Sm)						0.85	600 000	140 000	
Elongated single domain magnet (Fe50%Co50%)						0.905	80 000	40	mechanically weak

NOTE: The magnetic properties of materials depend critically on the manufacture and previous history of the specimen. The values in the tables above should therefore be taken as typical only.

DENSITY OF WATER (kg m^{-3}) AS A FUNCTION OF TEMPERATURE AT 1 ATMOSPHERE PRESSURE

Temperature $t/°C$	0	2	4	6	8	10	12	14	16	18
0	999·87	999·97	1000	999·97	999·88	999·73	999·52	999·27	998·97	998·62
20	998·23	997·80	997·32	996·81	996·26	995·67	995·05	994·40	993·71	992·99
40	992·2	991·5	990·7	989·8	989·0	988·1	987·2	986·2	985·3	984·3
60	983·2	982·2	981·1	980·1	978·9	977·8	976·7	975·5	974·3	973·1
80	971·8	970·6	969·3	968·0	966·7	965·3	964·0	962·6	961·2	959·8

Density at 100°C = 958·4; at 110°C = 951; at 150°C = 917; at 200°C = 863 kg m^{-3}.

NOTE: water has a maximum density at 3·98°C (277·13 K).

SATURATED PRESSURE AND SPECIFIC VOLUME OF WATER VAPOUR

Temp. $t/°C$	Temp. T/K	Saturated Vapour Pressure p_{sat}/MPa	Specific Volume V_c/m^3kg^{-1}	Temp. $t/°C$	Temp. T/K	Saturated Vapour Pressure p_{sat}/MPa	Specific Volume $V_c/m^3 kg^{-1}$
0	273·15	0·0006107	206·3	110	383	0·1433	1·2106
*0·01	273·15	0·0006112	206·1	120	393	0·1985	0·8920
1	274·15	0·0006565	192·6	130	403	0·2701	0·6685
2	275·15	0·0007054	179·9	140	413	0·3614	0·5088
3	276·15	0·0007575	168·2	150	423	0·4760	0·3926
4	277·15	0·0008129	157·3	160	433	0·6180	0·3068
5	278·15	0·0008719	147·1	170	443	0·7920	0·2426
8	281·15	0·0010721	121·0	180	453	1·0027	0·1938
10	283	0·001227	106·4	190	463	1·2552	0·1563
15	288	0·001704	77·97	200	473	1·555	0·1271
20	293	0·002337	57·84	220	493	2·320	0·08601
25	298	0·003166	43·40	240	513	3·348	0·05964
30	303	0·004242	32·93	260	533	4·694	0·04212
40	313	0·007375	19·55	280	553	6·419	0·03011
50	323	0·01234	12·04	300	573	8·592	0·02162
60	333	0·01992	7·678	320	593	11·29	0·01544
70	343	0·03116	5·045	340	613	14·61	0·01078
80	353	0·04736	3·408	360	633	18·67	0·006967
90	363	0·07011	2·361	374·14	**647·29	22·12	0·003155
100	373	0·101325	1·673				

*Triple point   **Critical point

SPECIFIC HEAT CAPACITY OF WATER AT 1 ATMOSPHERE PRESSURE

Temperature $t/°C$	Specific Heat Capacity $c_p/J kg^{-1} K^{-1}$	Temperature $t/°C$	Specific Heat Capacity $c_p/J kg^{-1} K^{-1}$	Temperature $t/°C$	Specific Heat Capacity $c_p/J kg^{-1} K^{-1}$
0	4217·4	35	4177·9	70	4189·3
5	4201·9	40	4178·3	75	4192·5
10	4191·9	45	4179·2	80	4196·1
15	4185·5	50	4180·4	85	4200·2
20	4181·6	55	4182·1	90	4204·8
25	4179·3	60	4184·1	95	4210·0
30	4178·2	65	4186·5	100	4215·7

Temperature $t/°C$	Vapour Pressure p_{sat}/mm Hg	Vapour Pressure p_{sat}/Pa	Density ρ/kg m^{-3}	Specific Heat Capacity C_p/J kg^{-1} K^{-1}	Temperature T/K
−20	$1·8 \times 10^{-5}$	$2·4 \times 10^{-3}$	13 644·56	140·31	253
0	$1·8 \times 10^{-4}$	$2·4 \times 10^{-2}$	13 595·08	139·67	273
20	$1·2 \times 10^{-3}$	0·16	13 545·88	139·08	293
40	$6·1 \times 10^{-3}$	0·81	13 496·95	138·53	313
60	0·025	3·3	13 448·25	138·02	333
80	0·089	11·9	13 399·77	137·56	353
100	0·273	36·4	13 351·48	137·13	373
120	0·746	99·5	13 303·4	136·76	393
140	1·85	247	13 255·4	136·42	413
160	4·19	559	13 207·5	136·13	433
180	8·80	1170	13 159·7	135·88	453
200	17·28	2304	13 112·0	135·67	473

RELATIVE HUMIDITIES FROM WET- AND DRY-BULB THERMOMETERS
(exposed in Standard Screen)

The relative humidity is defined as the ratio, expressed as a percentage, of the actual vapour pressure to the saturation vapour pressure at the temperature of the dry bulb. The dry bulb thermometer is an ordinary thermometer; the 'wet-bulb' thermometer is similar in design and has its bulb enclosed in a wick, the other end of which dips into water. By capillary action the thermometer bulb is wet and under the usually encountered conditions evaporation of the water lowers the temperature of the bulb. The difference in reading of the two thermometers is the 'Depression of the wet bulb'. The tables below give relative humidities for various values of the dry bulb temperature and the depression. Temperatures are in degrees Celsius.

Depression of Wet Bulb /°C	Dry Bulb Temperature/°C															
	0	2	4	6	8	10	12	14	16	18	20	22	24	26	28	30
0·5	91	92	93	93	94	94	95	95	95	95	96	96	96	96	96	96
1·0	81	84	85	86	87	88	89	90	90	91	91	92	92	92	93	93
1·5	73	76	78	80	81	82	83	85	85	86	87	87	88	88	89	89
2·0	64	68	71	73	75	77	78	79	81	82	83	83	84	85	85	86
2·5	55	61	64	66	69	71	73	75	76	77	78	80	80	81	82	83
3·0	46	52	57	60	63	66	68	70	71	73	74	76	77	78	78	79
3·5	38	45	49	54	57	60	63	65	67	69	70	72	73	74	75	76
4·0	29	37	43	48	51	55	58	60	63	65	66	68	69	71	72	73
4·5	21	29	36	41	46	50	53	56	58	61	63	64	66	67	69	70
5·0	13	22	29	35	40	44	48	51	54	57	59	61	62	64	65	67
5·5	5	14	22	29	35	39	43	47	50	53	55	57	59	61	62	64
6·0		7	16	24	29	34	39	42	46	49	51	54	56	58	59	61
6·5			9	17	24	29	34	38	42	45	48	50	53	54	56	58
7·0				11	19	24	29	34	38	41	44	47	49	51	53	55
7·5				5	14	20	25	30	34	38	41	44	46	49	51	52
8·0					8	15	21	26	30	34	37	40	43	46	48	50
8·5						10	16	22	26	30	34	37	40	43	45	47
9·0						6	12	18	23	27	31	34	37	40	42	44
9·5							8	14	19	23	28	31	34	37	40	42
10·0								10	15	20	24	28	31	34	37	39

INTERNATIONAL PRACTICAL TEMPERATURE SCALE 1968

Boiling and freezing temperatures listed below refer to standard atmospheric pressure of 101325 Pa except where stated otherwise.

	$t/°C$	T/K
Primary Reference Temperatures		
Equilibrium Hydrogen, triple point	−259·34	13·81
Equilibrium Hydrogen, boiling temperature at pressure 33330·6 Nm⁻² (25 mm Hg)	−256·108	17·042
Equilibrium Hydrogen, boiling temperature	−252·87	20·28
Neon, boiling temperature	−246·048	27·102
Oxygen, triple point	−218·789	54·361
Oxygen, boiling temperature	−182·962	90·188
Water, triple point	0·01	273·16
Water, boiling temperature	100·00	373·15
Zinc, freezing temperature	419·58	692·73
Silver, freezing temperature	961·93	1235·08
Gold, freezing temperature	1064·43	1337·58
Secondary Reference Temperatures		
Normal Hydrogen, triple point	−259·194	13·956
Normal Hydrogen, boiling temperature	−252·753	20·397
Neon, triple point	−248·595	24·555
Nitrogen, triple point	−210·002	63·148
Nitrogen, boiling temperature	−195·802	77·348
Carbon dioxide, sublimation point	−78·476	194·674
Mercury, freezing temperature	−38·862	234·288
Water, ice point	0	273·15
Phenoxybenzene, triple point	26·87	300·02
Benzoic acid, triple point	122·37	395·52
Indium, freezing temperature	156·634	429·784
Bismuth, freezing temperature	271·442	544·592
Cadmium, freezing temperature	321·108	594·258
Lead, freezing temperature	327·502	600·652
Mercury, boiling temperature	356·66	629·81
Sulphur, boiling temperature	444·674	717·824
Copper–aluminium eutectic, freezing temperature	548·23	821·38
Antimony, freezing temperature	630·47	903·89
Aluminium, freezing temperature	660·37	933·52
Copper, freezing temperature	1084·5	1357·6
Nickel, freezing temperature	1455	1728
Cobalt, freezing temperature	1494	1767
Palladium, freezing temperature	1554	1827
Platinum, freezing temperature	1772	2045
Rhodium, freezing temperature	1963	2236
Iridium, freezing temperature	2447	2720
Tungsten, freezing temperature	3387	3660

REFRACTIVE INDICES (n) AGAINST AIR, FOR THE MEAN SODIUM D
LINE (589·3 nm)

Calcite (ord)	1·658	Polystyrene	1·591
Calcite (extr)	1·486	Potassium alum	1·456
Canada balsam	1·530	Potassium iodide	1·667
Diamond	2·417	Quartz (ord)	1·544
Felspar	1·52	Quartz (extr)	1·553
Fluorspar	1·434	Rock salt (NaCl)	1·544
Glass, crown	1·48–1·61	Ruby	1·76
Glass, flint	1·53–1·96	Silver bromide	2·25
Ice	1·31	Sodium fluoride	1·326
Perspex	1·495	Sylvine (KCl)	1·490

WAVELENGTHS OF IMPORTANT SPECTRAL LINES IN AIR AT 15°C
AND 1 ATMOSPHERE PRESSURE. UNITS, nm (10^{-9} m)

Spectral line	Wavelength λ/nm	Spectral line	Wavelength λ/nm
K red	766·5	Fe and Ca green (E)	527·0
O red A	759·4	Mg green (b_1)	518·3
O red B	687·0	Mg green (b_2)	517·3
Li red	670·8	Mg green (b_4)	516·7
Hα red (c)	656·3	*Cd green	508·582
*Cd red	643·84696	Hβ blue-green (F)	486·1
Li orange	610·4	*Cd blue	479·991
Na orange (D_1)	589·59	Sr blue	460·7
Na orange (D_2)	589·00	Li blue	460·3
He yellow (D_3)	587·56	Hg blue	435·8
Hg yellow	579·0	Hγ blue (G_1)	434·0
Hg yellow	577·0	Fe and Ca blue (G)	430·8
Hg green	546·1	Ca blue (g)	422·7
Tl green	535·0	Hg and K violet	404·7

*Accepted standard lines

THE ELECTROMAGNETIC SPECTRUM

Type of radiation	Frequency ν/Hz	Wavelength λ/m	Wave No. σ/m^{-1}	Quantum Energy
		10^{-16}	10^{16}	12 400 MeV
	10^{24}			
		10^{-15}	10^{15}	1 240 MeV
	10^{23}			
		10^{-14}	10^{14}	124 MeV
	10^{22}			
		10^{-13}	10^{13}	12·4 MeV
gamma rays	10^{21}			
		10^{-12}	10^{12}	1·24 MeV
	10^{20}			
		10^{-11}	10^{11}	124 keV
	10^{19}			
		10^{-10}	10^{10}	12·4 keV
X-rays	10^{18}			
		10^{-9}	10^{9}	1·24 keV
	10^{17}			
		10^{-8}	10^{8}	124 eV
	10^{16}			
Violet		10^{-7}	10^{7}	12·4 eV
$\lambda \sim 4 \times 10^{-7}$m Ultra-violet	10^{15}			

Visible Spectrum

Type of radiation	Frequency ν/Hz	Wavelength λ/m	Wave No. σ/m^{-1}	Quantum Energy
Red Infra-red		10^{-6}	10^{6}	
$\lambda \sim 7 \times 10^{-7}$m	10^{14}			
		10^{-5}	10^{5}	
	10^{13}			
		10^{-4}	10^{4}	
	10^{12}			
		10^{-3}	10^{3}	
	10^{11}			
		10^{-2}	10^{2}	
Microwaves, radar	10^{10}			
		10^{-1}	10	
	10^{9}			
		1	1	
	10^{8}			
		10	10^{-1}	
	10^{7}			
		10^{2}	10^{-2}	
Short waves	10^{6}			
Long waves		10^{3}	10^{-3}	
	10^{5}			
		10^{4}	10^{-4}	

SPEED OF SOUND AT ROOM TEMPERATURE

Substance	Temp. $t/°C$	Speed $v/m\ s^{-1}$	Substance	Speed $v/m\ s^{-1}$
Air	0	331·3	Aluminium	5100
Hydrogen	0	1284	Brass	3500
Oxygen	0	316	Copper	3800
Water	25	1498	Iron	5000
Oak (along fibre)	15	3850	Lead	1200
Glass	20	5000	Mercury	1452

N.B.—The velocity of sound can vary according to the crystalline state and previous history of the specimen. The values quoted for solids are for longitudinal waves in thin specimens.

LOUDNESS OF SOUNDS

Intensity in terms of threshold-intensity I/I_{min}	Intensity I/dB	Loudness $L/phon$
1	0	Threshold of hearing
10	10 (1 bel)	Virtual silence
10^2	20	Quiet room
10^3	30	Watch ticking at 1 m
10^4	40	Quiet street
10^5	50	Quiet conversation
10^6	60	Quiet motor at 1 m
10^7	70	Loud conversation
10^8	80	Door slamming
10^9	90	Busy typing room
10^{10}	100	Near loud motor horn
10^{11}	110	Pneumatic drill
10^{12}	120	Near aeroplane engine
10^{13}	130	Threshold of pain

Limits of Audibility—Between 30 and 30 000 Hz (approximately).

MUSIC

The consonant frequency intervals.

Name	Octave	Fifth	Fourth	Major Third	Major Sixth	Minor Third	Minor Sixth
Frequency Ratio . . .	1 : 2	2 : 3	3 : 4	4 : 5	3 : 5	5 : 6	5 : 8

Musical Scales—Vibration Ratios

	C	D	E	F	G	A	B	C
Basic* Scale	24	27	30	32	36	40	45	48
	1·000	1·125	1·250	1·333	1·500	1·667	1·875	2·000
Intervals	..	$\frac{9}{8}$	$\frac{10}{9}$	$\frac{16}{15}$	$\frac{9}{8}$	$\frac{10}{9}$	$\frac{9}{8}$	$\frac{16}{15}$

*The Basic Scale is frequently referred to as the Natural or Diatonic Scale.

The vibration-numbers in the Basic Scale must bear the given ratios to each other, but their absolute values are matter of convention.

The London International Conference of May 1939 agreed that the international standard of concert pitch should be based on 440Hz for the treble A, *i.e.* 264 for the 'Middle C'.

In the EQUALLY TEMPERED SCALE the octaves remain as before, but 11 notes are introduced between them, the intervals being made equal and each $^{12}\sqrt{2}$, *i.e.* 1·0595, say 1·06 (approx.).

The following is such an equally tempered chromatic scale based on 440Hz as the treble A:

	Frequency ν/Hz		Frequency ν/Hz		Frequency ν/Hz
C′	261·6	F	349·2	A	440·0
C♯	277·2	F♯	370·0	A♯	466·2
D	293·7	G	392·0	B	493·9
D♯	311·1	G♯	415·3	C″	523·2
E	329·6				

ABSORPTION COEFFICIENTS OF BUILDING MATERIALS; UNIT, SABINE

	Frequency					
	125 Hz	250 Hz	500 Hz	1000 Hz	2000 Hz	4000 Hz
Acoustic plaster, 13 mm	0·15	0·20	0·35	0·60	0·60	0·50
Acoustic tiles, 20 mm	0·10	0·35	0·70	0·75	0·65	0·50
Brick, unglazed	0·03	0·03	0·03	0·04	0·05	0·07
Carpet, on concrete	0·02	0·06	0·14	0·37	0·60	0·65
Carpet with foam underlay	0·08	0·24	0·57	0·69	0·71	0·73
Curtain, heavy velour	0·14	0·35	0·55	0·72	0·70	0·65
Linoleum, on concrete	0·02	0·03	0·03	0·03	0·03	0·02
Glass, heavy plate	0·18	0·06	0·04	0·03	0·02	0·02
Glass, window	0·35	0·25	0·18	0·12	0·07	0·04
Plaster	0·013	0·015	0·02	0·03	0·04	0·05
Plywood panelling, 10 mm	0·28	0·22	0·17	0·09	0·10	0·11
Polystyrene, expanded, 13 mm	0·05	0·15	0·40	0·35	0·20	0·20
Polyurethane foam, 50 mm	0·25	0·50	0·85	0·95	0·90	0·90
Tiles, glazed	0·01	0·01	0·01	0·01	0·02	0·02
Wood parquet	0·04	0·04	0·07	0·06	0·06	0·07

TIME

1 mean solar second = $\dfrac{1}{86\,400}$ of a mean solar day.

1 sidereal day = 86 164·090 6 mean solar seconds.
1 tropical (civil) year = 365·242 mean solar days = $3·155\,692\,597\,47 \times 10^7$s
1 sidereal year = 365·256 mean solar days.
1 mean synodical or lunar month = 29·531 mean solar days.

N.B.—Centuries are not leap years unless divisible by 400.

DISTANCE

1 Astronomical Unit (AU) = mean sun–earth distance = $1·495\,985(5) \times 10^{11}$m
1 Parsec (pc) = $3·085\,6(1) \times 10^{16}$ m = $2·062\,648 \times 10^5$ AU = 3·261 5 ly
1 Light year (ly) = $9·460\,5 \times 10^{15}$ m = $6·324 \times 10^4$ AU = 0·3066 pc

THE SUN

Radius = $6·960 \times 10^8$ m = $4·326 \times 10^5$ miles
Surface area = $6·087 \times 10^{18}$ m²
Volume = $1·412 \times 10^{27}$ m³
Mass = $1·99 \times 10^{30}$ kg
Mean density = 1409 kg m⁻³
Rate of energy production = $3·90 \times 10^{26}$W
Gravity at surface = 274 m s⁻²
Moment of inertia = $6·0 \times 10^{46}$ kg m²
Escape velocity at surface = 618 km s⁻¹
Sidereal period of rotation = 25·38 days
Period of rotation with respect to the earth = 27·28 days $\Big\}$ latitude 16°

THE MOON

Radius = 1738 km = 1080 miles
Surface area = $3·796 \times 10^{13}$ m²
Volume = $2·199 \times 10^{19}$ m³
Mass = $7·349 \times 10^{22}$ kg = 1/81.4 × mass of earth
Mean density = 3340 kg m⁻³
Sidereal period of moon about earth = 27·32 mean solar days
Mean synodical or lunar month = 29·531 mean solar days
Mean distance from the earth = $3·844 \times 10^8$ m = $2·39 \times 10^5$ miles
Surface area of the moon at some time visible from the earth = 59%
Gravity at surface = 1·62 m s⁻²
Moment of inertia = $8·8 \times 10^{28}$ kg m²
Escape velocity at surface = 2·38 km s⁻¹

THE SOLAR SYSTEM

Body	Equatorial radius R/m	Mass M/kg	Density ρ/kg m^{-3}	Distance from Sun d/m	Surface gravity g/m s^{-2}	Ellipticity e	Eccentricity of orbit ε	Inclination to ecliptic i/°	No. of satellites N	Sidereal period T_s	Rotational period T_r
Sun	$6{\cdot}960 \times 10^{8}$	$1{\cdot}989 \times 10^{30}$	1409		274	0			—	—	25-38d
Moon	$1{\cdot}738 \times 10^{6}$	$7{\cdot}353 \times 10^{22}$	3340	$1{\cdot}496 \times 10^{11}$	1·62	—	0·055	5·144	—	27·32d	27·32d
Mercury	$2{\cdot}42 \times 10^{6}$	$3{\cdot}301 \times 10^{23}$	5420	$5{\cdot}791 \times 10^{10}$	3·76	0	0·2056	7·004	0	87·97d	58·7d
Venus	$6{\cdot}085 \times 10^{6}$	$4{\cdot}869 \times 10^{24}$	5250	$1{\cdot}082 \times 10^{11}$	8·77	0	0·0068	3·394	0	224·7d	243d
Earth	$6{\cdot}378 \times 10^{6}$	$5{\cdot}978 \times 10^{24}$	5510	$1{\cdot}496 \times 10^{11}$	9·81	0·0034	0·0167	0	1	365·3d	23·93h
Mars	$3{\cdot}375 \times 10^{6}$	$6{\cdot}420 \times 10^{23}$	3960	$2{\cdot}279 \times 10^{11}$	3·80	0·007	0·0934	1·850	2	687d	24·6h
Jupiter	$7{\cdot}14 \times 10^{7}$	$1{\cdot}899 \times 10^{27}$	1330	$7{\cdot}783 \times 10^{11}$	24·9	0·062	0·0481	1·306	12	11·86a	9·9h
Saturn	$6{\cdot}04 \times 10^{7}$	$5{\cdot}685 \times 10^{26}$	680	$1{\cdot}427 \times 10^{12}$	10·4	0·096	0·0533	2·489	10	29·46a	10·2h
Uranus	$2{\cdot}36 \times 10^{7}$	$8{\cdot}686 \times 10^{25}$	1600	$2{\cdot}869 \times 10^{12}$	10·4	0·06	0·0507	0·773	5	84·02a	10·7h
Neptune	$2{\cdot}23 \times 10^{7}$	$1{\cdot}025 \times 10^{26}$	1650	$4{\cdot}498 \times 10^{12}$	13·8	0·02	0·0040	1·773	2	164·8a	15·8h
Pluto	3×10^{6}	5×10^{23}	3000	$5{\cdot}900 \times 10^{12}$	4	—	0·2533	17·142	0	248a	6·3d

Notes: Ellipticity of a planet is defined by $(R_e - R_p)/R_e$, where R_e is the equatorial radius and R_p is the polar radius. The sidereal period of a planet is the time to move once round its orbit. Periods are measured in hours (h), days (d) or years (a).

This scale is used to indicate the brightness of a star as observed by the human eye. A visual magnitude of 6 is just visible to the human eye, and brighter stars are indicated by *smaller* visual magnitudes on a logarithmic scale. A change in visual magnitude of 1 unit indicates a change in the brightness of the star by a factor $\sqrt[5]{100} = 2 \cdot 512$. Thus a star of magnitude 1 is 100 times brighter than a star of magnitude 6 and a star of magnitude -1 is $2 \cdot 512$ times brighter than a star of magnitude 0.

THE BRIGHTEST STARS
in decreasing order of brightness

Star	Visual Magnitude	Distance, $d/10^{15}$m	Distance d/light years
α Canis Majoris (Sirius)	−1·6	82	8·7
α Carinae (Canopus)	−0·9	1700	180
α Centauri (Rigil Kent)	0·1	41	4·3
α Lyrae (Vega)	0·1	251	27
α Boötes (Arcturus)	0·2	340	36
α Aurigae (Capella)	0·2	420	44
β Orionis (Rigel)	0·3	11 000	1200
α Canis Minoris (Procyon)	0·5	107	11
α Eridani (Achernar)	0·6	1300	140
β Centauri (Hadar)	0·9	1900	200

APPROXIMATE GALACTIC DISTANCES
including Baade's correction
(M = Messier Catalogue No.   N.G.C. = new general catalogue No.)

Great Nebula in Andromeda (M31, N.G.C. 224)	22×10^5 light years	$= 210 \times 10^{20}$ m
Nebula in Andromeda (M32)	15×10^5 „	$= 140 \times 10^{20}$ m
Nebula in Andromeda (N.G.C. 205)	15×10^5 „	$= 140 \times 10^{20}$ m
Nebula in Triangulum (M33)	15×10^5 „	$= 140 \times 10^{20}$ m
Large Magellanic Cloud (in Dorado)	$1 \cdot 4 \times 10^5$ „	$= 13 \times 10^{20}$ m
Small Magellanic Cloud (in Toucan)	$1 \cdot 5 \times 10^5$ „	$= 14 \times 10^{20}$ m
Crab Nebula (N.G.C. 1952)	6×10^3 „	$= 0 \cdot 6 \times 10^{20}$ m

THE EARTH

Polar radius = 6356·8 km
Equatorial radius = 6378·2 km
Mean radius = 6371 km = 3960 miles
Surface area = $5·101 \times 10^{14}$ m²
Volume = $1·083 \times 10^{21}$ m³
Mass = $5·977 \times 10^{24}$ kg
Mean density = 5517 kg m⁻³
Mean distance to the sun (AU) = $1·496 \times 10^{11}$ m = $9·2868 \times 10^7$ miles
Distance to sun at perihelion = $1·471 \times 10^{11}$ m = $9·136 \times 10^7$ miles
Distance to sun at aphelion = $1·521 \times 10^{11}$ m = $9·447 \times 10^7$ miles
Gravity at surface = 9·80665 m s⁻² (standard)
Moment of inertia about axis of rotation = $8·04 \times 10^{37}$ kg m²
Escape velocity at surface = 11·2 km s⁻¹
Rotational velocity at equator = 465 m s⁻¹
Mean Velocity in its orbit about the sun = 29·78 km s⁻¹
Solar constant = solar energy incident on unit area normal to the sun's rays at
the earth's mean distance, per unit time = 1400 J m⁻² s⁻¹

1° of latitude at equator = 110·5 km = 68·70 miles.
1° of latitude at poles = 111·7 m = 69·41 miles
1° of longitude at equator = 111·3 km = 69·17 miles.
Inclination of equator to ecliptic = 23° 27′.
Greatest height (Mt. Everest) = 8847·7 m = 29 028 ft (1954 Indian Survey).
Greatest depth (Marianas Trench) = 11 033 m = 35 960 ft.
Land area = $148·8 \times 10^6$ km² = $5·747 \times 10^7$ miles².
Ocean area = $361·3 \times 10^6$ km² = $13·95 \times 10^7$ miles².

COMPOSITION OF THE ATMOSPHERE

The composition of dry air is remarkably constant all over the globe and through-
out the entire troposphere. The proportions *by volume* of the various components
are given below (after A. F. Paneth, 1939, 1952).

Substance	% by volume	Substance	% by volume
N_2	78·09	CH_4	$2·0 \times 10^{-4}$
O_2	20·95	Kr	1×10^{-4}
Ar	0·93	H_2	5×10^{-5}
*CO_2	0·03	N_2O	5×10^{-5}
Ne	$1·8 \times 10^{-3}$	Xe	9×10^{-6}
He	$5·2 \times 10^{-4}$	Rn	6×10^{-18}

*This varies somewhat near towns and industrial areas.

THE ICAO STANDARD ATMOSPHERE

The International Civil Aviation Organization have defined a standard atmosphere which is an attempt to represent atmospheric conditions in temperate latitudes. At sea level, standard pressure and acceleration of gravity are assumed for a temperature of 288 K (15°C). The air is assumed to be a perfect gas of fixed composition.

Sea level properties of the ICAO atmosphere

Collision frequency	$6 \cdot 9204 \times 10^9$ s^{-1}	Pressure	$1 \cdot 01325 \times 10^5$ Pa
Density	$1 \cdot 225$ kg m^{-3}	Scale height	$8 \cdot 4344 \times 10^3$ m
Gravitational acceleration	$9 \cdot 80665$ m s^{-2}	Speed of sound	$340 \cdot 29$ m s^{-1}
Kinematic viscosity	$1 \cdot 4607 \times 10^{-5}$ m^2 s^{-1}	Temperature	$288 \cdot 15$ K
Mean free path	$6 \cdot 6317 \times 10^{-8}$ m	Thermal conductivity	$2 \cdot 5339 \times 10^{-2}$
Molar volume	$2 \cdot 3645 \times 10^{-2}$ m^3 mol^{-1}		W m^{-1} K^{-1}
Molecular weight	$28 \cdot 966$	Viscosity	$1 \cdot 7894 \times 10^{-5}$
Number density	$2 \cdot 5475 \times 10^{25}$ m^{-3}		kg m^{-1} s^{-1}
Particle speed	$4 \cdot 5894 \times 10^2$ m s^{-1}		

Variation of pressure, density and temperature with height

Geometric Height h/m	Pressure p/Pa	Density ρ/kg m^{-3}	Temp. T/K	Geometric Height h/m	Pressure p/Pa	Density ρ/kg m^{-3}	Temp. T/K
-250	104365	$1 \cdot 2547$	$289 \cdot 775$	6000	$47217 \cdot 6$	$0 \cdot 66011$	$249 \cdot 187$
0	101325	$1 \cdot 2250$	$288 \cdot 150$	7000	$41105 \cdot 2$	$0 \cdot 59002$	$242 \cdot 700$
$+250$	$98357 \cdot 6$	$1 \cdot 1959$	$286 \cdot 525$	8000	$35651 \cdot 6$	$0 \cdot 52579$	$236 \cdot 215$
500	$95461 \cdot 2$	$1 \cdot 1673$	$284 \cdot 900$	9000	$30800 \cdot 7$	$0 \cdot 46706$	$229 \cdot 733$
750	$92634 \cdot 6$	$1 \cdot 1392$	$283 \cdot 276$	10000	$26499 \cdot 9$	$0 \cdot 41351$	$223 \cdot 252$
1000	$89876 \cdot 2$	$1 \cdot 1117$	$281 \cdot 651$	15000	$12111 \cdot 8$	$0 \cdot 19475$	$216 \cdot 650$
1500	$84559 \cdot 6$	$1 \cdot 0581$	$278 \cdot 402$	20000	$5529 \cdot 3$	$0 \cdot 08891$	$216 \cdot 650$
2000	$79501 \cdot 4$	$1 \cdot 0066$	$275 \cdot 154$	25000	$2594 \cdot 2$	$0 \cdot 04008$	$221 \cdot 552$
2500	$74691 \cdot 7$	$0 \cdot 95695$	$271 \cdot 906$	30000	$1197 \cdot 0$	$0 \cdot 01841$	$226 \cdot 509$
3000	$70121 \cdot 1$	$0 \cdot 90925$	$268 \cdot 659$	32000	$889 \cdot 1$	$0 \cdot 01355$	$228 \cdot 490$
3500	$65780 \cdot 3$	$0 \cdot 86340$	$265 \cdot 413$	50 000	$80 \cdot 96$	$1 \cdot 041 \times 10^{-3}$	271
4000	$61660 \cdot 4$	$0 \cdot 81935$	$262 \cdot 166$	100 000	$3 \cdot 095 \times 10^{-2}$	$5 \cdot 062 \times 10^{-7}$	213
5000	$54048 \cdot 2$	$0 \cdot 73643$	$255 \cdot 676$	200 000	$8 \cdot 806 \times 10^{-5}$	$2 \cdot 56 \times 10^{-10}$	1198

NOTE: the above table is reproduced by permission of the International Civil Aviation Organization, Montreal. The last three sets of values in this table are taken from the COSPAR International Reference Atmosphere, 1965 (CIRA 1965) by permission of the publishers, North Holland Publishing Co., Amsterdam.

Principal Elements in Earth's Crust (% by mass)

Oxygen 49·13%, Silicon 26·0%, Aluminium 7·45%, Iron 4·2%, Calcium 3·25%, Sodium 2·4%, Potassium 2·35%, Magnesium 2·35%, Hydrogen 1%. All others 1·87%.

Principal Elements in the Hydrosphere (% by mass)

Oxygen 85·89%, Hydrogen 10·82%, Chlorine 1·90%, Sodium 1·06%. All others 0·33%.

ACCELERATION OF GRAVITY (g)

At a latitude, λ, and height, h (measured in metres), above sea-level, the acceleration of gravity is given by the expression:

$$g/\mathrm{m\,s^{-2}} = 9{\cdot}80616 - 0{\cdot}025928\cos2\lambda + 0{\cdot}000069\cos^2 2\lambda - 0{\cdot}000003\,h$$

Geophysical data for various places of importance.

In the following tables, values of the acceleration of gravity and the length of the seconds pendulum are calculated using the formula above. In addition, magnetic data, calculated for the year 1970 are included. These have been obtained from the International Reference Geomagnetic Field and excluding local variations should not be in error by more than 1%. Declination is positive Eastward and Angle of Dip positive downwards. Magnetic Induction for geophysical fields is often measured in *gammas*. Where 1 gamma $= 10^{-9}$ Tesla or 1 gamma $= 1$ nT.

Location	Position		Acceleration of Gravity $g/\mathrm{m\,s^{-2}}$	Length of Seconds Pendulum l/m	Declination $D/°$	Horizontal Component of Earth's Magnetic Field $H/\mathrm{Am^{-1}}$	Horizontal Component of Earth's Magnetic Induction B_H/nT	Angle of Dip $I/°$
Equator	0°0′		9·78030	0·99094				
Madras	13°5′N	80°18′E	9·78281	0·99120	−2·23	32·4	40660	9·1
Calcutta	22°35′N	88°21′E	9·7882	0·99175	−0·82	31·2	39200	29·9
Sydney	33°55′S	151°10′E	9·7968	0·99262	11·9	20·2	25380	−64·1
Capetown	33°56′S	18°28′E	9·7966	0·99260	−24·6	9·86	12390	−65·1
Tokyo	35°40′N	139°45′E	9·79801	0·99275	−6·4	24·3	30530	48·3
New York	40°40′N	73°50′W	9·80267	0·99322	−11·6	14·6	18420	71·0
Paris	48°52′N	2°20′E	9·80943	0·99390	−5·5	16·0	20140	64·7
London	51°25′N	0°20′W	9·81183	0·99415	−7·0	15·0	18820	66·8
Edinburgh	55°57′N	3°13′W	9·8158	0·99455	−9·4	13·2	16620	70·1
Leningrad	59°55′N	30°25′E	9·81929	0·99490	7·0	12·1	15180	72·8
N'th Pole	90°0′N		9·8322	0·99621				

TABLE OF ENERGY EQUIVALENTS

Energy associated with:	Basic equation	J	eV	calorie	kWh
1 Joule (J)		1	$6{\cdot}242 \times 10^{18}$	0·2389	$2{\cdot}778 \times 10^{-7}$
1 eV	$E = eV$	$1{\cdot}602 \times 10^{-19}$	1	$3{\cdot}828 \times 10^{-20}$	$4{\cdot}450 \times 10^{-26}$
1 calorie		4·186	$2{\cdot}613 \times 10^{17}$	1	$1{\cdot}163 \times 10^{-6}$
1 kilowatt-hour (kWH)		$3{\cdot}600 \times 10^{6}$	$2{\cdot}247 \times 10^{25}$	$8{\cdot}600 \times 10^{5}$	1
1 kilogram (kg)	$E = mc^2$	$8{\cdot}988 \times 10^{16}$	$5{\cdot}610 \times 10^{35}$	$2{\cdot}147 \times 10^{16}$	$2{\cdot}497 \times 10^{10}$
1 electron mass (m_e)	$E = mc^2$	$8{\cdot}187 \times 10^{-14}$	$5{\cdot}110 \times 10^{5}$	$1{\cdot}956 \times 10^{-14}$	$2{\cdot}274 \times 10^{-20}$
1 unified mass unit (u)	$E = mc^2$	$1{\cdot}492 \times 10^{-10}$	$9{\cdot}313 \times 10^{8}$	$3{\cdot}564 \times 10^{-11}$	$4{\cdot}144 \times 10^{-17}$
1 Hertz (Hz)	$E = h\nu$	$6{\cdot}626 \times 10^{-34}$	$4{\cdot}136 \times 10^{-15}$	$1{\cdot}583 \times 10^{-14}$	$1{\cdot}841 \times 10^{-40}$
1 reciprocal metre	$E = hc/\lambda$	$1{\cdot}986 \times 10^{-25}$	$1{\cdot}240 \times 10^{-6}$	$4{\cdot}745 \times 10^{-26}$	$5{\cdot}517 \times 10^{-32}$
1 Kelvin (K)	$E = kT$	$1{\cdot}381 \times 10^{-23}$	$8{\cdot}620 \times 10^{-5}$	$3{\cdot}299 \times 10^{-24}$	$3{\cdot}836 \times 10^{-30}$

There are various relationships, basic to physics, which introduce the energy associated with a system. Of these, the following are of especial importance:

Einstein's equation, $E = mc^2$,
Planck's equation, $E = h\nu$,
Boltzmann's equation, $E = kT$.

Partly as a result of the importance of the concept of energy, there are many different units in which it is measured. The cgs unit is the erg, the SI unit is the Joule, while in atomic and nuclear physics, it is always measured in electron volts (eV). Other units in common use are the calorie and the kilowatt hour. The table below is based on the equations above and may be used for converting most of the commonly encountered energy units. It gives equivalent quantities in horizontal lines. Thus $1 \text{ kg} = 8 \cdot 988 \times 10^{16} \text{ J} = 1 \cdot 097 \times 10^{30}$ electron masses etc.

TABLE OF ENERGY EQUIVALENTS (CONT.)

kg	me	u	Hz	m^{-1}	K
$1 \cdot 113 \times 10^{-17}$	$1 \cdot 221 \times 10^{13}$	$6 \cdot 702 \times 10^{9}$	$1 \cdot 509 \times 10^{33}$	$5 \cdot 034 \times 10^{24}$	$7 \cdot 244 \times 10^{22}$
$1 \cdot 783 \times 10^{-36}$	$1 \cdot 956 \times 10^{-6}$	$1 \cdot 074 \times 10^{-9}$	$2 \cdot 418 \times 10^{14}$	$8 \cdot 066 \times 10^{5}$	$1 \cdot 160 \times 10^{4}$
$4 \cdot 658 \times 10^{-17}$	$5 \cdot 110 \times 10^{13}$	$2 \cdot 805 \times 10^{10}$	$6 \cdot 316 \times 10^{33}$	$2 \cdot 107 \times 10^{25}$	$3 \cdot 032 \times 10^{23}$
$4 \cdot 007 \times 10^{-11}$	$4 \cdot 396 \times 10^{19}$	$2 \cdot 413 \times 10^{16}$	$5 \cdot 432 \times 10^{39}$	$1 \cdot 812 \times 10^{31}$	$2 \cdot 608 \times 10^{29}$
1	$1 \cdot 097 \times 10^{30}$	$6 \cdot 024 \times 10^{26}$	$1 \cdot 356 \times 10^{50}$	$4 \cdot 525 \times 10^{41}$	$6 \cdot 511 \times 10^{39}$
$9 \cdot 112 \times 10^{-31}$	1	$5 \cdot 487 \times 10^{-4}$	$1 \cdot 235 \times 10^{20}$	$4 \cdot 121 \times 10^{11}$	$5 \cdot 931 \times 10^{9}$
$1 \cdot 661 \times 10^{-27}$	$1 \cdot 822 \times 10^{3}$	1	$2 \cdot 251 \times 10^{23}$	$7 \cdot 511 \times 10^{14}$	$1 \cdot 081 \times 10^{13}$
$7 \cdot 375 \times 10^{-51}$	$8 \cdot 090 \times 10^{21}$	$4 \cdot 441 \times 10^{-24}$	1	$3 \cdot 336 \times 10^{-9}$	$4 \cdot 800 \times 10^{-11}$
$2 \cdot 210 \times 10^{-42}$	$2 \cdot 425 \times 10^{-12}$	$1 \cdot 331 \times 10^{-15}$	$2 \cdot 997 \times 10^{8}$	1	$1 \cdot 439 \times 10^{-2}$
$1 \cdot 537 \times 10^{-40}$	$1 \cdot 686 \times 10^{-10}$	$9 \cdot 255 \times 10^{-14}$	$2 \cdot 084 \times 10^{10}$	$6 \cdot 952 \times 10^{1}$	1

The most common unit of radioactivity is the Curie (Ci). Originally defined as the volume of radon gas in equilibrium with 1 g radium, it has since become associated with the number of disintegrations occurring per second in 1 g of radium free from its daughter products viz. 3.7×10^{10} disintegrations per second. In modern usage, the curie has been redefined to agree with this result, and other units have been introduced as given below.

One *curie* (Ci) of any radioactive substance is that quantity in which 3.7×10^{10} atoms disintegrate per second. The millicurie (mCi) and microcurie (μCi) are in common usage.

The *rutherford* is the unit of activity corresponding to 10^6 disintegrations per second. Thus 37 rutherford = 1 mCi.

The *roentgen* (r) was originally suggested as a unit of radiation and has become of universal use in defining the quantities of X-rays or γ-rays present. In 1937, the Fifth International Congress of Radiobiology recommended the following definition:

The *roentgen* is that quantity of X- or γ- radiation such that the associated corpuscular emission per 0·001293 g of dry air produces, in air, ions carrying 1 esu of quantity of electricity of either sign. (*N.B.* this mass of air occupies 1 cm³ at STP).

Dose rates are often measured in units of roentgen hour^{-1} or milliroentgen hour^{-1} (mr h^{-1})

The *rad* is defined as the absorbed dose of radiation when 1 g of material absorbs 100 ergs of energy. 1 rad = 10^{-2} J kg^{-1}.

The *roentgen equivalent man* (rem) is the unit Dose Equivalent used in Radiation Protection. The Dose Equivalent is the product of the Absorbed Dose (measured in rad) and the quality factor Q, of the radiation. The value of Q indicates how damaging the particular radiation is, compared with 200 keV X-rays. Thus, low energy β-rays have $Q = 1.7$, while neutrons impinging on the eye have $Q = 30$.

A useful, but approximate formula for calculation of dose rates from γ-ray point sources is

$$\text{Dose rate (r hr}^{-1}) \simeq (5000 \, C \, E)/d^2$$

where C is the activity of the source in curies, E the energy of the γ-ray emitted in MeV and d is the distance from the source in cm. If more than one γ-ray is emitted, the total dose rate is the sum of the individual dose rates.

The naturally radioactive materials with the exception of a few isotopes, e.g. ^{40}K are the heavy elements of atomic number $Z > 80$. Three 'families' are known in which one substance decays to another which in turn continues the process until a stable material (lead) is attained. The decay process involves the emission of an electron (β-particle) or an α-particle from the nucleus. In the former case, the mass number, A, remains unchanged while Z increases by unity, while the latter emission involves a decrease in A of four and a decrease in Z of two as the α-particle is the helium nucleus. In any one 'family' the mass numbers alter in steps of four only. In the Thorium family each value of A can be described by the number ($4n$), the Uranium family by ($4n+2$) and the Actinium family by ($4n+3$). The apparently missing family ($4n+1$) has been found as a result of the artificial production of heavy isotopes. It does not appear naturally because the longest half life is short compared with the age of the earth.

The law of radioactive decay

All radioactive substances transform at a rate which is proportional to the number of atoms present. If there are N_0 atoms present at the zero of time, then at time, t, there are N_t, where

$$N_t = N_0 \exp -(\lambda t)$$

Here, λ, is a constant for the particular type of atom considered and is known as the transformation constant. The rate at which an atom decays is often measured in terms of the mean lifetime of the atom, τ, or the half-value period, $T_{\frac{1}{2}}$, which is often abbreviated to the half-life. The relation between these constants is:

$$\tau = 1/\lambda = T_{\frac{1}{2}}/\log_e 2$$

For values of the half-value periods of important isotopes see section 27, Table of Isotopes, P87ff.

24 Properties of Inorganic Compounds

In the following table, properties refer to room temperature, 293 K. Enthalpies of Formation refer to the substance in the crystalline (c), liquid (lq), or gaseous (g) states at 293 K. A negative value indicates that heat is evolved in the formation of the compound, while a positive value indicates absorption of heat. The following abbreviations are used:

bl.	black	effl.	efflorescent	s.	sublimes
col.	colourless	ex.	explodes	tetr.	tetragonal
crys.	crystals	gn.	green	trig.	trigonal
cub.	cubic	hex.	hexagonal	visc.	viscous
d.	dissociates	mono.	monoclinic	w.	white
delq.	deliquescent	rh.	rhombic	yel.	yellow

Formula		Molecular Weight M/g mol^{-1}	Melting Point T_M/K	Boiling Point T_B/K	Density ρ/kg m^{-3}	Refractive Index n	Enthalpy of Formation ΔH^{θ}/kJ mol^{-1}		Description
Al	Al$_2$O$_3$	101·96	2290	3250	3965	1·768	−1670	c	Corundum, w. trig.
Ag	AgBr	187·78	705	1600 (d)	6473	2·252	−99·5	c	pale yel. cub
	AgCl	143·32	728	1820	5560	2·071	−127	c	w. cub
	AgNO$_3$	169·87	485	717 (d)	4352	1·744	−123	c	col. rh.
As	AsBr$_3$	314·65	306	494	3540		−195·0	c	col. prisms
	AsCl$_3$	181·28	265	403	2163		−335	lq	Oily liquid
	As$_2$O$_3$	197·84	588		3738	1·755	−1310	c	col. cub. (As$_4$O$_6$)
Au	AuCl$_3$	303·33	527 (d)		3900		−118	c	red delq.
Ba	BaCl$_2$	208·25	1240	1820	3856	1·736	−860·1	c	col. mono.
	BaO	153·34	2196	2300	5720	1·98	−558·1	c	col. cub.
Be	BeCl$_2$	79·92	678	790	1899	1·719	−511·7	c	w. delq. needles
	BeO	25·01	2800	4170	3010	1·719	−610·9	c	w. hex.
C	CO	28·01	74	84	1·25		−110·5	g	col. gas
	CO$_2$	44·01	162	195	1·98		−393·5	g	col. gas
Ca	CaCO$_3$	100·09	1612	d	2930	1·6809	−1206·9	c	Aragonite, col. rh.
	CaCl$_2$	110·99	1045	1900	2150	1·52	−795·0	c	w. delq. cub.
	CaO	56·08	2850	3120	3300	1·837	−635·5	c	col. cub.

	Formula	Molecular Weight M/g mol⁻¹	Melting Point T_M/K	Boiling Point T_B/K	Density ρ/kg m⁻³	Refractive Index n	Enthalpy of Formation $\Delta H_f^\ominus$/kJ mol⁻¹		Description
Cd	CdBr₂	272·22	840	1136	5192		−314·4	c	w. effl. needles
	CdCl₂	183·32	841	1233	4047		−389·1	c	w. cub.
	CdO	128·40	1200 (d)		8150		−254·6	c	brown cub.
Co	CoCl₂	129·84	997*	1322	2940		−326	c	blue crys. *in HCl gas
	CoO	74·93	2208		6450		−239	c	brown cub.
	Co(OH)₂	92·95	d		3597		−548·9	c	rose-red rh.
Cs	CsCl	168·36	919	1560	3988	1·534	−433·0	c	col. delq. cub.
Cu	CuO	79·54	1599		6400		−155·2	c	bl. cub. or trig.
	CuSO₄	223·14			3605	1·733	−769·9	c	gn/w. rh.
	CuSO₄·5H₂O	249·68			2284	1·537	−2278	c	blue trig.
	Cu₂O	143·08	1508		6000	2·705	−166·7	c	red cub.
Fe	FeS	87·91	1470	d	4740		−95·1	c	blue hex.
	Fe₂O₃	159·69	1838		5240	3·042	−822·2	c	red or bl. trig
	Fe₃O₄	231·54	1810 (d)		5180	2·42	−1117	c	bl. cub.
H	HBr	80·92	185	206	3·5		−36·2	g	col. gas
	HCl	36·46	158	188	1·0		−92·3	g	col. gas
	HF	20·01	190	293	0·99		−268·6	g	col. gas
	HI	127·91	222	238	5·66		+25·9	g	col. gas
	HNO₃	63·01	231	356	1503		−173·2	lq	col. liquid
	H₂O	18·02	273	373	1000	1·333	−285·9	lq	col. liquid
	H₂SO₄	98·08	284	610	1841		−814·0	lq	col. visc. liquid
Hg	HgCl	236·05	670 (s)		7150	1·973	−265*	c	w. tetr. (*Hg₂Cl₂)
	HgCl₂	271·50	549	575	5440	1·859	−230	c	w. rh.
	HgO	216·59	800 (d)		11100	2·5	−90·4	c	yel. or red rh.
K	KCl	74·56	1049	1770 (s)	1984	1·490	−435·9	c	col. cub.
	KHCO₃	100·12	400 (d)		2170	1·482	−959·4	c	mono.
	K₂CO₃	138·21	1164	d	2428	1·531	−1146·1	c	w. delq.
	K₂O	94·20	620 (d)		2320		−361·5	c	w. cub.
Li	LiCl	42·39	887	1600	2068	1·662	−408·8	c	w. delq. cub.
Mg	MgBr₂	184·13	970		3720		−517·6	c	w. delq.
	MgCO₃	84·32	620	1200	2958	1·700	−1112	c	w. trig.
	MgCl₂	95·22	981	1685	2320	1·675	−641·8	c	col. hex.
	MgF₂	62·31	1539	2512		1·378	−1102	c	col. tetr.
	MgH₂	26·33	550 (d)						w. tetr.
	MgI	278·12	1000 (d)		4430		−359	c	w. delq.
	MgO	40·31	3100	3900	3580	1·736	−601·8	c	col. cub.
	Mg(OH)₂	58·33	620		2360	1·562	−924·7	c	w. trig.
	MgSO₄	120·37	1397		2660	1·56	−1278	c	col. rh.
Mn	MnO	70·94			5440	2·16	−385	c	gray/gn cub.
	MnO₂	86·94	808 (d)		5026		−520·9	c	bl. rh.
	MnO₃	102·94							red delq.
	Mn₂O₃	157·87	1350 (d)		4500		−971·1	c	brown/bl. cub.
	Mn₂O₇	221·87	279	328 (d)	2396				red oil
	Mn₃O₄	228·81	1978		4856	2·46	−1386	c	brown/bl. tetr.
N	NH₃	17·03	195	240	0·77		−46·2	g	col. gas
	NH₄Cl	53·49	613 (s)		1527	1·64	−315·4	c	w. cub.
	NO	30·01	110	121	1·34		+90·4	g	col. gas
	NO₂	44·01	182	185	1·98		+33·8	g	red/brown gas (N₂O₄)
	N₂O₃	76·01	171	277 (d)	1447		+83·8	g	red/brown gas

Formula		Molecular Weight M/g mol^{-1}	Melting Point T_M/K	Boiling Point T_B/K	Density ρ/kg m^{-3}	Refractive Index n	Enthalpy of Formation ΔH_f^θ/kJ mol^{-1}		Description
Na	NaBr	102.90	1028	1660	3203	1.641	−359.9	c	col. cub.
	NaCl	58.44	1074	1686	2165	1.544	−411.0	c	col. cub.
	NaF	41.99	1261	1968	2558	1.326	−569	c	col. tetr.
	NaH	24.00	1100 (d)		920	1.470	−57.3	c	silver needles
	NaHCO$_3$	84.00	540 (d)		2159	1.500	−947.7	c	w. mono. powder
	NaHSO$_4$	120.06	590		2435		−1126	c	col. tricl.
	NaI	149.89	924	1577	3667	1.774	−288.0	c	col. cub.
	NaOH	40.00	592	1660	2130		−426.7	c	w. delq.
	Na$_2$CO$_3$	105.99	1124	d	2532	1.535	−1131	c	w. powder
	Na$_2$O	61.98	1548 (s)		2270		−416	c	w/gray delq.
	Na$_2$SO$_4$	142.04			2680	1.477	−1384	c	mono (→hex at 510 K)
Ni	NiCl$_2$	129.62	1274		3550		−316	c	yel. delq.
	NiO	74.71	2260		6670	2.37	−244	c	gn/bl. cub.
P	PCl$_3$	137.33	161	349	1574	1.503	−320	c	col. fuming liquid
	PCl$_5$	208.24		435 (s)	4.65		−463.2	g	delq. tetr.
	PH$_3$	34.00	140	185			+5.2	g	col. gas
	P$_2$O$_3$	109.95	297	447*	2135		−820	lq	w. delq. mono. * in N$_2$
	P$_2$O$_4$	125.95	370	450*	2540				w delq. rh. *in vacuo
	P$_2$O$_5$	141.94	850	875	2390		−3012*		w. delq. amor. *P$_4$O$_{10}$
Pb	PbCl$_2$	278.10	774	1220	5850	2.217	−359.2	c	w. rh
	PbCl$_4$	349.00	258	378 (ex)	3180				yel. liquid
	PbO	223.19	1161		9530		−219.2	c	red amor.
	PbO$_2$	239.19	560 (d)		9375	2.229	−276.6	c	brown tetr.
	PbS	239.25	1387		7500	3.912	−100.4	c	lead gray cub.
	Pb$_3$O$_4$	685.57	770 (d)		9100		−718.4	c	red amor.
Rb	RbCl	120.92	988	1660	2800	1.494	−430.5	c	cub.
S	SO$_2$	64.06	200	263	2.93		−296.9	g	col. gas
	SO$_3$	80.06	306	318	1927*		−395.2	g	col. gas (*liquid)
Sb	SbBr$_3$	361.48	370	550	4148	1.74	−260	c	col. rh.
	SbCl$_3$	228.11	347	556	3140		−382	c	col. rh. delq.
	SbCl$_5$	299.02	276	352	2336		−438	lq	pale yel. liquid
Si	SiC	40.10	3000		3217	2.654	−111.7	c	blue/bl. trig.
	SiCl$_4$	169.90	203	331	1483	1.412	−640.2	lq	col. fuming liquid
	SiH$_4$	32.12	88	161	1.44		+34	g	col. gas
	SiO	44.09	1975	2150	2130				w. cub.
	SiO$_2$	60.08	1880	2500	1.544		−911	c	Quartz, hex.
Sn	SnCl$_4$	260.50	240	387	2226		−511.3	lq	col. fuming liquid
	SnO	134.69	1350 (d)		6446		−286	c	bl. cub.
	SnO$_2$	150.69	1400	2100 (s)	6950	1.997	−581	c	w. tetr.
Sr	SrCl$_2$	158.53	1146	1520	3052	1.536	−828	c	w. rh.
	SrO	103.62	2700	3300	4700	1.870	−590	c	col. cub.
Ti	TiCl$_4$	189.71	248	409	1726		−750	lq	col. liquid
	TiO$_2$	79.90	2098		4170	2.586	−912	c	bl. rh.
U	UC$_2$	262.05	2650	4640	11280		−176	c	metallic crystals
	UO$_2$	270.03	2800		10960		−1130	c	bl. rh.
W	WC	195.86	3140	6300	15630		−38.0	c	gray. cub. powder
	WO$_3$	231.85	1746		7160		−840	c	yel. rh.
Zn	ZnCO$_3$	125.39	570 (d)		4398	1.818	−813	c	w. trig.
	ZnCl$_2$	136.28	556	1005	2910	1.687	−416	c	w. delq.
	ZnO	81.37	2100		5606	2.004	−348	c	w. hex.

25 Properties of Organic Compounds (at 293K)

Enthalpies of Formation refer to the substance in the crystalline (c), liquid (lq), or gaseous (g) states at 293 K. A negative value indicates evolution of heat during formation of the compound, while a positive value indicates absorption of heat. Enthalpy changes on combustion refer to combustion at a pressure of 1 atmosphere and temperature 293 K, the final products being liquid water, and gaseous carbon dioxide and nitrogen.

Name and Formula	Molecular Weight	Melting Point T_M/K	Boiling Point T_B/K	Density ρ/kg m^{-3}	Refractive Index, n	Enthalpy of Formation ΔH_f^θ/kJ mol^{-1}	Heat of Combustion ΔH_c/kJ mol^{-1}	Alternative Name
Hydrocarbons								
Methane CH$_4$	16·04	91	109			−74·85 g	890·4g	
Ethane C$_2$H$_6$	30·07	90	185			−84·7 g	1560 g	
Propane C$_3$H$_8$	44·11	83	231			−103·8 g	2220 g	
n-Butane n-C$_4$H$_{10}$	58·13	135	273	579	1·3543	−146·2 lq	2877 g	
2-Methyl propane iso-C$_4$H$_{10}$	58·13	114	261	557		−134·6 g	2869 g	Isobutane
n-Pentane n-C$_5$H$_{12}$	72·15	143	309	626	1·3575	−173 lq	3509 lq	
n-Hexane n-C$_6$H$_{14}$	86·18	178	342	660	1·3751	−198·8 lq	4163 lq	
n-Heptane n-C$_7$H$_{16}$	100·21	183	372	638	1·3878	−224·4 lq	4853 lq	
n-Octane n-C$_8$H$_{18}$	114·23	216	399	702	1·3974	−250 lq	5512 lq	
Ethene n-C$_2$H$_4$	28·05	104	169	1·26		+52·3 g	1411 g	Ethylene
Propene C$_3$H$_6$	42·08	88	226	519	1·3567	+20·4 g	2059 g	Propylene
Ethyne C$_2$H$_2$	26·04	192	189	618		+226·7 g	1300 g	Acetylene
Benzene C$_6$H$_6$	78·12	279	353	879	1·5011	+48·7 lq	3273 lq	
Cyclohexane C$_6$H$_{12}$	84·16	280	354	779	1·4266	−156·2 lq	3924 lq	
Halogen derivatives of hydrocarbons								
Monochloromethane CH$_3$Cl	50·49	175	249	916		−81·9 g	687 g	Methyl chloride
Dichloromethane CH$_2$Cl$_2$	84·93	178	313	1327	1·4242	−117 lq	447 g	Methylene dichloride
Trichloromethane CHCl$_3$	119·38	210	335	1483	1·4459	−132 lq	373 lq	Chloroform
Tetrachloromethane CCl$_4$	153·82	250	350	1594	1·4601	−139·5 lq	156 lq	Carbon tetrachloride
Bromomethane CH$_3$Br	94·94	180	277	1676	1·4218	−35·6 g	770 g	Methyl bromide
Iodomethane CH$_3$I	141·94	207	316	2279	1·5380	−8·4 lq	815 lq	Methyl iodide
Alcohols								
Methanol CH$_3$OH	32·04	179	338	791	1·3288	−238·7 lq	726·5 lq	
Ethanol C$_2$H$_5$OH	46·07	156	352	789	1·3611	−277·7 lq	1371 lq	
n-Propanol n-C$_3$H$_7$OH	60·11	147	371	803	1·3850	−300 lq	2017 lq	
Propane-1,2,3-triol C$_3$H$_8$O$_3$	92·11	293	d	1261	1·4746	−103·9 lq	1661 lq	Glycerol
Acids								
Ethanoic acid CH$_3$COOH	60·05	290	391	1049	1·3716	−488·3 lq	876 lq	Acetic acid
Propanoic acid C$_2$H$_5$COOH	74·08	252	414	993	1·3869	−509 lq	1527 lq	Propionic acid
n-Butanoic acid n-C$_3$H$_7$COOH	88·12	269	437	958	1·3980	−538·9 lq	2194 lq	
Benzoic acid C$_6$H$_5$COOH	122·13	396	522	1266	1·504	−390 c	3227 c	
Miscellaneous								
Ethanal CH$_3$CHO	44·05	152	294	783	1·3316	−166·4 g	1192 lq	Acetaldehyde
2-Propanone CH$_3$·CO·CH$_3$	58·08	178	329	790	1·3588	−216·7 g	1821 g	Acetone
Methoxymethane CH$_3$·O·CH$_3$	46·07	135	250			−185 g	1454 g	Dimethylether
Ethoxyethane C$_2$H$_5$·O·C$_2$H$_5$	74·12	157	308	714	1·3526	−279·6 lq	2761 lq	Diethylether
Urea CO(NH$_2$)$_2$	60·06	408	d	1323	1·484	−333·2 c	634 c	
Glycine NH$_2$·CH$_2$·COOH	75·07	d		828		−528·6 c	981 c	

A	α	Alpha	I	ι	Iota	P	ρ	Rho
B	β	Beta	K	κ	Kappa	Σ	σ	Sigma
Γ	γ	Gamma	Λ	λ	Lambda	T	τ	Tau
Δ	δ	Delta	M	μ	Mu	Y	υ	Upsilon
E	ε	Epsilon	N	ν	Nu	Φ	ϕ	Phi
Z	ζ	Zeta	Ξ	ξ	Xi	X	χ	Chi
H	η	Eta	O	o	Omicron	Ψ	ψ	Psi
Θ	θ	Theta	Π	π	Pi	Ω	ω	Omega

27 Table of Isotopes

The following table lists all the stable isotopes and also includes a selection of important unstable isotopes.

Column 1 gives the atomic number, symbol and mass number of the isotope. The mass numbers of stable isotopes are printed in bold type. An asterisk with the mass number indicates an isomer (metastable excited nucleus). Column 2 gives the abundance, a, of the isotope in the naturally occurring element and for the unstable isotopes indicates the type of decay by the symbols: α, $\beta-$, $\beta+$, -radiation, p proton emission, n neutron emission, k electron capture, i.t. isomeric transition with emission of γ-rays. Column 3 gives the atomic masses in unified mass units. The masses of the nuclei can be obtained from these by subtraction of the masses of the Z electrons of mass 0.000549 u each.

Column 4 gives for unstable isotopes the maximum energy, E, of the emitted particles for several possible disintegrations in the order shown in column 2. Column 5 gives the corresponding half-value periods in seconds (s), minutes (min), days (d) or years (a). Column 6 gives the inner quantum number of the nucleus and the energy of gamma-rays (MeV); column 7 gives the nuclear magnetic moment (nuclear magnetons).

Element Z	A	a [%] or disint.	M u	E MeV	T	I or $E\gamma$	μ
−1e	−	stable	$0.000\,548_9$				
1p	1	stable	$1.007\,276$				
0n	1	$\beta-$	$1.008\,665_4$	0.78	10.8 min	$\frac{1}{2}$	$-1.913\,1$
1 H	1	99.985	$1.007\,825_2$			$\frac{1}{2}$	$+2.792\,6$
D	2	0.015	$2.014\,102_2$			I	$+0.857\,3$
T	3	$\beta-$	$3.016\,049$	0.018	12.3 a	$\frac{1}{2}$	$+2.978\,5$
2 He	3	1.4×10^{-4}	$3.016\,030$			$\frac{1}{2}$	$-2.127\,4$
	4	~100	$4.002\,604$			0	0
	5	n	$5.012\,3_0$		$\sim6 \times 10^{-20}$ s		
	6	$\beta-$	$6.018\,9$	3.5	0.82 s	0	0
3 Li	5	p	$5.012\,5$			$\frac{3}{2}$	
	6	7.42	$6.015\,12_6$			I	$+0.821\,9$
	7	92.58	$7.016\,00_5$			$\frac{3}{2}$	$+3.256$
	8	$\beta-$	$8.022\,48_8$	~13	0.84 s		1.65
	9	$\beta-+n$	9.02_7	$\beta\sim8$	0.17 s		

Table of Isotopes

Element Z	A	a[%] or disint.	M u	E MeV	T	I or E_γ	μ
4 Be	7	K	$7 \cdot 016\ 93_1$		53 d	0·48	
	8	2α	$8 \cdot 005\ 30_8$	0·05	$\sim 3 \times 10^{-16}$s	0	0
	9	100	$9 \cdot 012\ 18_6$			$\frac{3}{2}$	$-1 \cdot 1774$
	10	β⁻	$10 \cdot 013\ 54_5$	0·56	$2 \cdot 7 \times 10^6$a	0	0
	11	β⁻	$11 \cdot 021\ 6_6$	$\underline{11 \cdot 5}$; 9·3	14 s	2...8	
5 B	8	β⁺+2α	$8 \cdot 024\ 61_2$	β 14	0·8 s		
	9		$9 \cdot 013\ 33_5$				
	10	19·6	$10 \cdot 012\ 93_9$			3	$+1 \cdot 801$
	11	80·4	$11 \cdot 009\ 305$			$\frac{3}{2}$	$+2 \cdot 689$
	12	β⁻(+α)	$12 \cdot 014\ 35_3$	13·4	0·02 s	1	
	13	β⁻	$13 \cdot 017\ 78$		0·04 s		
6 C	10	β⁺	$10 \cdot 016\ 8_1$	2·1	19 s	0	0
	11	β⁺	$11 \cdot 011\ 43$	0·96	20·5 min	$\frac{3}{2}$	
	12	98·89	12 (Stand.)			0	0
	13	1·11	$13 \cdot 003\ 35_4$			$\frac{1}{2}$	$+0 \cdot 7022$
	14	β⁻	$14 \cdot 003\ 242$	0·158	5570 a	0	0
	15	β⁻	$15 \cdot 010\ 60_0$	9·8; $\underline{4 \cdot 5}$	2·3 s	5·3	
7 N	12	β⁺ (β⁺+3a)	$12 \cdot 018\ 7$	$\underline{16 \cdot 6}$; 12·2	0·012 s		
	13	β⁺	$13 \cdot 005\ 73_9$	1·2	10·1 min	$\frac{1}{2}$	
	14	99·63₄	$14 \cdot 003\ 074$			1	$+0 \cdot 4037$
	15	0·36₆	$15 \cdot 000\ 10_8$			$\frac{1}{2}$	$-0 \cdot 2830$
	16	β⁻	$16 \cdot 006\ 0_9$	10·4; 4·3	7·4 s	2	
	17	β⁻+n	$17 \cdot 008\ 4_5$	3·7 (0·9)	4·1 s		
8 O	14	β⁺	$14 \cdot 008\ 597$	1·8	72 s	2·3	
	15	β⁺	$15 \cdot 003\ 07_2$	1·68	124 s	$\frac{1}{2}$; noγ	
	16	99·76	$15 \cdot 994\ 915_1$			0	0
	17	0·037 4	$16 \cdot 999\ 13_3$			$\frac{5}{2}$	$-1 \cdot 893$
	18	0·204	$17 \cdot 999\ 160$			0	0
	19	β⁻	$19 \cdot 003\ 58$	4·6; $\underline{3 \cdot 2}$	29·4 s		
9 F	17	β⁺	$17 \cdot 002\ 10$	$1 \cdot 75$	66 s		
	18	β⁺(K)	$18 \cdot 000\ 95$	0·65	110 min		
	19	100	$18 \cdot 998\ 40$			$\frac{1}{2}$	$+2 \cdot 628$
	20	β⁻	$19 \cdot 999\ 99$	5·4	11 s	2; 1·63	
10 Ne	18	β⁺	$18 \cdot 005\ 7_2$	3·4	1·3 s		
	19	β⁺	$19 \cdot 001\ 89$	2·2	18 s		
	20	90·92	$19 \cdot 992\ 440$			0	0
	21	0·257	$20 \cdot 993\ 84_9$			$\frac{3}{2}$	$-0 \cdot 662$
	22	8·82	$21 \cdot 991\ 384$			0	0
	23	β⁻	$22 \cdot 994\ 47$	$\underline{4 \cdot 4}$; 3·9	38 s	0·44	
	24	β⁻	$23 \cdot 993\ 6$	$\underline{2 \cdot 0}$; 1·1	3·4 min	0·47; 0·88	
11 Na	20	β⁺+α	$20 \cdot 008_9$	$3 \cdot 5 < E_\beta <$ $7 \cdot 3\ E_\alpha > 2$	0·3 s		
	21	β⁺	$20 \cdot 997\ 6_4$	2·5	23 s		
	22	β⁺(K)	$21 \cdot 994\ 44$	0·54	2·6 a	3; 1·28	$+1 \cdot 746$
	23	100	$22 \cdot 989\ 77_3$			$\frac{3}{2}$	$+2 \cdot 2165$
	24	β⁻	$23 \cdot 990\ 97$	1·4	15 h	4;1·37; 2·75	$+1 \cdot 688$
	*24	i. t.,β⁻		~6	0·02 s	0·47	
	25	β⁻	$24 \cdot 989_9$	$\underline{3 \cdot 8}$; 2·8	60 s	0·4...1·6	

Element Z	A	$a[\%]$ or disint.	M u	E MeV	T	I or E_γ	μ
12 Mg	23	β^+	$22.994\,14$	$\underline{3.1}$; 2.6	12 s	0.44	
	24	78.7	$23.985\,04_5$			0	0
	25	10.1	$24.985\,84_0$			$\frac{5}{2}$	-0.855
	26	11.2	$25.982\,59_1$			0	0
	27	β^-	$26.984\,35$	$\underline{1.8}$; 1.6	9.5 min	0.83; 1.0	
	28	β^-	$27.983\,8_8$	0.4	21.3 h	$0.03\ldots1.3$	0
13 Al	24	$\beta^+(+\alpha)$	24.000	$\beta{:}8.5$; $\alpha{:}2$	2.1 s	$1.4\ldots7.1$	
	25	β^+	$24.990\,4_1$	3.2	7.2 s	1.6	
	26	$\beta^+(K)$	$25.986\,90$	1.2	7×10^5 a	1.1; 1.8	
	*26	β^+		3.2	6.5 s		
	27	100	$26.981\,53_5$			$\frac{5}{2}$	$+3.639$
	28	β^-	$27.981\,91$	2.9	2.3 min	1.8	
	29	β^-	$28.980\,44$	$\underline{2.5}$; 1.5	6.6 min	1.3; 2.4	
14 Si	27	β^+	$26.986\,70$	3.8	4 s		
	28	92.21	$27.976\,93$			0	0
	29	4.70	$28.976\,49$			$\frac{1}{2}$	$-0.554\,8$
	30	3.09	$29.973\,76$			0	0
	31	β^-	$30.975\,35$	1.5	157 min		
	32	β^-	$31.974\,0$	~0.1	~700 a	0	0
15 P	28	β^+	27.992	$\underline{10.6}$	0.28 s	$1.8\ldots7.6$	
	29	β^+	$28.981\,8_2$	3.9	4.3 s		
	30	β^+	$29.978\,3_2$	3.2	2.5 min		
	31	100	$30.973\,76_3$			$\frac{1}{2}$	$+1.131$
	32	β^-	$31.973\,90_8$	1.7	14.5 d	I; noγ	-0.252
	33	β^-	$32.971\,73$	0.25	25 d		
	34	β^-	33.973_3	$\underline{5.1}$; 3.2	12.4 s	2.1	
16 S	31	β^+	$30.979\,6_0$	4.4	2.6 s		
	32	95.0	$31.972\,07_4$			0	0
	33	0.76	$32.971\,46$			$\frac{3}{2}$	$+0.643$
	34	4.22	$33.967\,86$			0	0
	35	β^-	$34.969\,03$	0.167	87 d	$\frac{3}{2}$	$(+)\,1.0$
	36	0.014	$35.967\,0_9$			0	0
	37	β^-	36.971_0	4.7; $\underline{1.6}$	5.0 min	2.7	
	38	β^-	37.971_2	3.0; $\underline{1.1}$	2.9 h	1.9	
17 Cl	32	$\beta^+(+\alpha)$	31.986	$\underline{9.5}$; 8.2	0.3 s	4.3; 4.8	
	33	β^+	32.977_4	4.5	2.8 s		
	34	β^+	$33.973\,7_6$	4.5	1.5 s		
	*34	i. t.,β^+		2.5; 1.3	32.4 min	$\begin{cases}\text{i. t.; }0.14\\ \gamma{:}1.2\ldots3.3\end{cases}$	
	35	75.5	$34.968\,85$			$\frac{3}{2}$	$+0.821$
	36	$\beta^-(K)$	$35.968\,3_1$	0.71	3×10^5 a	2	$+1.284$
	37	24.5	$36.965\,90$			$\frac{3}{2}$	$+0.684$
	38	β^-	$37.968\,0_0$	$\underline{4.8}$; 2.8; 1.1	37.3 min	1.6; 2.2	
	39	β^-	$38.968\,0$	$\underline{3.5}$; 2.2; $\underline{1.9}$	56 min	$0.2\ldots1.5$	
	40	β^-	39.970	7.5; 3.2	1.4 min	$1.5\ldots6.0$	
18 Ar	35	β^+	$34.975\,3$	4.96	2 s	1.2; 1.7	
	36	0.337	$35.967\,55$			0	0
	37	K	$36.966\,77$		34 d	$\frac{3}{2}$	1.0
	38	0.063	$37.965\,72$			0	0

Element Z	A	a[%] or disint.	M u	E MeV	T	I or E$_\gamma$	μ
	39	β⁻	38·964 3$_2$	0·56	265 a		
	40	99·60	39·962 38$_4$			0	0
	41	β⁻	40·964 5$_0$	2·5; 1·2	110 min	1·3	
19 K	37	β⁺	36·973 4	5	1·2 s		
	38	β⁺	37·969 1	2·7	7·7 min	2·2	
	*38	β⁺		5·1	0·95 s		
	39	93·10	38·963 71			$\frac{3}{2}$	+0·391
	40	0·011 8β⁻ K	39·964 01	1·32	1·3×10⁹ a	4	−1·297
				1·46			
	41	6·88	40·961 83			$\frac{3}{2}$	+0·215
	42	β⁻	41·962 4	3·6; 2·2	12·5 h	2; 1·5	−1·14
20 Ca	39	β⁺	38·970 7	5·5	0·9 s		
	40	96·97	39·962 59			0	0
	42	0·64	41·958 63			0	0
	43	0·14	42·958 78			$\frac{7}{2}$	−1·315
	44	2·1	43·955 49			0	0
	45	β⁻	44·956 19	0·26	165 d		
	46	0·003	45·953 6$_9$			0	0
	47	β⁻	46·954 5	1·94; 0·66	4·7 d	1·3	
	48	0·18	47·952 3$_6$			0	0
	49	β⁻	48·955 6$_6$	2·0; 0·9	8·8 min	3·1; 4·0	
21 Sc	45	100	44·955 92			$\frac{7}{2}$	+4·749
	46	β⁻	45·955 1$_7$	0·36	84 d		
	*46	i. t.			20 s	0·14	
	47	β⁻	46·952 4$_0$	0·6; 0·44	3·4 d	0·16	
	48	β⁻	47·952 2$_3$	0·65	44 h	1·0; 1·3	
22 Ti	46	7·93	45·952 63			0	0
	47	7·28	46·951 7$_6$			$\frac{5}{2}$	−0·787
	48	73·94	47·947 95			0	0
	49	5·51	48·947 87			$\frac{7}{2}$	−1·102
	50	5·34	49·944 79			0	0
	51	β⁻	50·946 6	2·1; 1·5	5·8 min	0·3; 0·6; 0·9	
23 V	48	β⁺, K	47·952 2$_6$	0·70	16 d	1·0; 1·3	
	50	0·24; K	49·947 17		4×10¹⁴ a	6	+3·341
	51	99·76	50·943 98			$\frac{7}{2}$	+5·14
	52	β⁻	51·944 8$_0$	2·6	3·77 min	1·4	
24 Cr	50	4·31	49·946 05			0	0
	51	K	50·944 79		28 d	$\frac{7}{2}$; 0·32	
	52	83·76	51·940 51			0	0
	53	9·55	52·940 65			$\frac{3}{2}$	−0·474
	54	2·38	53·938 88			0	0
	55	β⁻	54·941$_1$	2·8	3·5 min		
25 Mn	54	K	53·940 3$_6$		280 d	3; 0·84	3·3
	55	100	54·938 05			$\frac{5}{2}$	+3·462
	56	β⁻	55·938 9$_1$	2·9; 1·0	2·6 h	3; 0·8...2·1	+3·240
26 Fe	54	5·82	53·939 6$_2$			0	0
	55	K	54·938 30		2·7 a	0·21	
	56	91·66	55·934 9$_3$			0	0
	57	2·19	56·935 3$_9$			$\frac{1}{2}$	+0·09

Element Z	A	a[%] or disint.	M u	E MeV	T	I or Eγ	μ
	58	0·33	57·933 2₇			0	0
	59	β⁻	58·934 8₇	0·46; 0·27	45 d	1·1; 1·3	
27 Co	57	K	56·936 29		270 d	7/2	4·6
	58	K(β⁺)	57·935 7₅	0·47	71 d	2; 0·01...1·2	4·1
	59	100	58·933 19			7/2	+4·64
	60	β⁻	59·933 8₁	0·314	5·29 a	5; 1·17; 1·33	+3·8
	*60	i. t.,−		1·5	10·5 min	0·06	
28 Ni	58	67·9	57·935 3₄				0
	59	K	58·934 3₄		~10⁵ a		
	60	26·2	59·930 7₈			0	0
	61	1·2	60·931 0₅			3/2	0·3
	62	3·7	61·928 3₅			0	0
	63	β⁻	62·926₇	0·067	120 a		
	64	1·1	63·927 9₆			0	0
	65	β⁻	64·930 0₄	2·1; 1·0; 0·6	2·6 h	0·4; 1·1; 1·5	
29 Cu	63	69·1	62·929 5₉			3/2	+2·221
	64	β⁻, β⁺, K	63·929 7₆	β⁻0·57 β⁺0·66	12·8 h	1; 1·34	0·22
	65	30·9	64·927 7₉			3/2	+2·379
	66	β⁻	65·928 8₇	2·6; 1·6	5·1 min	1·0	
30 Zn	64	48·89	63·929 15			0	0
	65	K, β⁺	64·929 2₃	0·33	245 d	1·1	
	66	27·81	65·926 0₅			0	0
	67	4·11	66·927 1₅			5/2	+0·874
	68	18·57	67·924 8₇			0	0
	69	β⁻	68·926 7	0·9	55 min		
	*69	i. t.			14 h	0·44	
	70	0·62	69·925 3₅			0	0
	71	β⁻	70·928₀	2·3	2·2 min		
31 Ga	69	60·4	68·925 7			3/2	+2·011
	70	β⁻	69·926 0₅	1·6	21 min	1	
	71	39·6	70·924 8			5/2	+2·555
	72	β⁻	71·926 0	3·2...0·6	14 h	3; 0·5...2·8	−0·132
32 Ge	70	20·5	69·924 2₈			0	0
	71	K	70·925 1		11 d	1/2	0
	72	27·4	71·921 7			0	0
	73	7·8	72·923₄			9/2	−0·877
	74	36·5	73·921₂			0	0
	75	β⁻	74·922₈	1·2; 0·9	84 min	0·26	
	*75	i. t.			46 s	0·14	
	76	7·8	75·921₄			0	
	77	β⁻	76·923₆	2·2; 1·4; 0·7	12 h	0·22 u.a.	0
33 As	73	K	72·923₈		76 d	0·01; 0·05	
	74	K, β⁻, β⁺	73·923₉	β⁻: 1·4; 0·7 β⁺: 1·5; 0·9	18 d	0·6 u. a.	
	75	100	74·921			3/2	+1·435
	76	β⁻	75·922 4	3·0; 2·4	26·5 h	2; 0·6; 1·2	−0·90
	77	β⁻	76·920₇	0·7	39 h	0·09...0·5	

Element Z	A	a[%] or disint.	M u	E MeV	T	I or E$_\gamma$	μ
34 Se	74	0·9	73·922$_4$			0	0
	75	K	74·922$_5$		120 d	$\frac{5}{2}$; 0·02...0·4	+1·1
	76	9·0	75·919 2			0	0
	77	7·6	76·919 1			$\frac{1}{2}$	0·533
	*77	i. t.			17·4 s	0·16	
	78	23·5	77·917 4			0	0
	80	49·8	79·916 5$_1$			0	0
	81	β⁻		1·4	18 min		
	*81	i. t.			60 min	0·10	
	82	9·2	81·916$_7$			0	0
	83	β⁻	82·918$_9$	1·6;1·0;0·4	25 min	0·2...2·3	
	*83	β⁻		<u>3·4</u>; 1·5	70 s	0·4...2·0	
35 Br	79	50·54	78·918 4			$\frac{3}{2}$	+2·099
	80	β⁻(K, β⁺)		<u>2·0</u>; 1·4 (0·9)	18 min	I; 0·62	
	*80	i. t.			4·7 h	5; 0·05	
	81	49·46	80·916 3			$\frac{3}{2}$	+2·263
	82	β⁻	81·916 8$_0$	0·44	36 h	5; 0·5...1·5	(+)1·626
	87	β⁻, β⁻+n	86·922	β⁻: 8·0; <u>2·6</u>	56 s	5·4; 3	
36 Kr	78	0·35	77·920 3$_7$			0	0
	79	K(β⁺)	78·920 1	(0·6; 0·3)	34·5 h	0·04...0·8	0
	80	2·27	79·916 3$_9$			0	0
	82	11·56	81·913 4$_8$			0	0
	83	11·55	82·914 1$_3$			$\frac{9}{2}$	-0·97
	84	56·9	83·911 5$_0$			0	0
	85	β⁻	84·912$_4$	0·67	10 a	$\frac{9}{2}$; 0·5	-1·0
	*85	β⁻, i. t.		0·8	4·5 h	0·15	
	86	17·37	85·910 6$_2$			0	0
	87	β⁻	86·913$_4$	<u>3·8</u>; 1·3	78 min	0·4; 0·9; 2·6	
37 Rb	85	72·15	84·911$_7$			$\frac{5}{2}$	+1·348
	86	β⁻	85·911$_2$	<u>1·8</u>; 0·7	18·7 d	2; 1·1	-1·67
	*86	i. t.			1·0 min	0·6	
	87	27·85 β⁻	86·909$_2$	0·27	4·7×10¹⁰ a	$\frac{3}{2}$	+2·741
	88	β⁻	87·911$_2$	<u>5·2</u>; 3·6; 2·5	18 min	2; 0·9; 1·8	
38 Sr	84	0·56	83·913 3$_8$			0	0
	86	9·9	85·909$_3$			0	0
	87	7·0 (β⁻?)	86·908$_9$			$\frac{9}{2}$	-1·09
	*87	i. t.			2·9 h	0·39	
	88	82·6	87·905$_6$			0	0
	89	β⁻	88·907	1·46	51 d		
	90	β⁻→⁹⁰Y	89·907 3	0·54	28 a	no γ	
39 Y	89	100	88·905$_4$			$\frac{1}{2}$	-0·137
	90	β⁻	89·906$_7$	2·26	64 h	2; no γ	-1·6
	91	β⁻	90·906$_9$	1·5; 0·3	58 d	1·2	
	*91	i. t.			50 min	0·55	
40 Zr	90	51·5	89·904$_3$			0	0
	91	11·2	90·905$_3$			$\frac{5}{2}$	-1·30
	92	17·1	91·904$_6$			0	0
	94	17·4	93·906			0	0
	95	β⁻	94·908	<u>0·40</u>; 0·36	65 d	0·73; 0·76	

Element Z	A	a[%] or disint.	M u	E MeV	T	I or E_γ	μ
	96	2·8 (β⁻?)	95·908		>10¹⁷ a	0	0
	97	β⁻	96·911	1·9; 0·4	17·0 h	0·5 ... 2·6	
41 Nb	93	100	92·906₀			9/2	6·14
	*94	i. t.				0·042	
	95	β⁻	94·907	(0·9); 0·16	35 d	0·75; 0·77	
	*95	i. t.			90 h	0·23	
42 Mo	92	15·8	91·906₃			0	0
	94	9·0	93·904₇			0	0
	95	15·7	94·906			5/2	−0·910
	96	16·5	95·905			0	0
	97	9·5	96·906			5/2	−0·929
	98	23·5	97·906			0	0
	99	β⁻	98·908	1·2; 0·5	67 h	0·04 ... 0·78	
	100	9·6	99·908			0	0
	101	β⁻	100·908₉	2·2 ... 0·6	14·6 min	0·08 ... 2·1	
43 Tc	96	K	95·908		4·35 d	1·12	
	97	K			2·6×10⁶ a		
	99	β⁻	98·906	0·3	2·1×10⁵ a	9/2	+5·657
	*99	i. t.			6 h	0·14	
	101	β⁻	100·905₉	1·3 (1·1)	14 min	0·1 ... 0·9	
44 Ru	96	5·5	95·908			0	0
	97	K			2·9 d	0·4; 0·22; 0·33	
	98	1·9	97·906			0	0
	99	12·7	98·906			5/2	−0·6
	100	12·6	99·903₀			0	0
	101	17·1	100·904₁			5/2	−0·7
	102	31·6	101·903₇			0	0
	103	β⁻	102·905₆	0·2; 0·1	40 d	0·50; 0·61	
	104	18·6	103·905₅			0	0
	105	β⁻	104·907₃	1·15	4·45 h	0·26 ... 0·96	
	106	β⁻	105·907₀	0·04	1·0 a		
45 Rh	103	100	102·904₈			1/2	−0·088
	104	β⁻	103·906₂	2·4; 1·9;0·7	42 s	0·56; 1·2	
	*104	i. t. (β⁻)			4·4 min	0·05; 0·08	
	105	β⁻	104·905₃	0·56; 0·25	35 h	0·32 u. a.	
	*105	i. t.			30 s	0·13	
	106	β⁻	105·907₀	3·5; 3·1; 2·4	30 s	0·5 ... 2·7	
46 Pd	102	1·0	101·904₉			0	0
	103	K	102·905₄		17 d	0·06 ... 0·50	
	104	11·0	103·903₆			0	0
	105	22·2	104·904₆			5/2	−0·6
	106	27·3	105·903₂			0	0
	108	2·67	107·903₉			0	0
	109	β⁻	108·905₉	1·0	13·6 h		
	*109	i. t.			4·8 min	0·18	
	110	11·8	109·904₅			0	0
	111	β⁻	110·907₆	2·1	22 min	0·4 ... 1·4	
47 Ag	105	K	104·906₈		45 d	1/2; 0·06...0·65	
	107	51·4	106·905₀			1/2	−0·113
	*107	i. t.			44 s	0·094	

Element Z	A	a[%] or disint.	M u	E MeV	T	I or E$_\gamma$	μ
	108	β⁻(K, β⁺)	107·905₉	β⁻: 1·77; β⁺: 0·8	2·4 min	0·4; 0·6	
	109	48·6	108·904₇			½	−0·130
	*109	i. t.			39 s	0·088	
	110	β⁻	109·906₁	2·9; 2·2	24·5 s	0·66	
	*110	β⁻(i. t.)		0·53; <u>0·09</u>	253 d	6; 0·1 ... 1·5	
	111	β⁻	110·905₂	<u>1·05</u>; 0·7	7·5 d	½; 0·34	0·14
48 Cd	106	1·22	105·906₀			0	0
	108	0·88	107·904₀			0	0
	109	K	108·904₉		470 d	5/2; ¹⁰⁹Ag*	−0·83
	110	12·39	109·903₀			0	0
	111	12·75	110·904₂			½	−0·592
	112	24·07	111·902₈			0	0
	113	12·26(β⁻?)	112·904₆		≥3×10¹⁵ a	½	−0·619
	114	28·86	113·903₆			0	0
	115	β⁻	114·905₆	<u>1·1</u>; 0·6	2·3 d	0·2 ... 0·5	
	*115	β⁻		1·6	43 d	0·5 ... 1·3	
	116	7·58	115·905₀			0	0
	117	β⁻	116·907₄	1·8	50 min	0·42	
	*117	β⁻		1·0	3 h	0·3 ... 2·2	
49 In	113	4·3	112·904₃			9/2	+5·50
	*113	i. t.			1·7 h	½; 0·39	
	114	β⁻(K,β⁺)	113·905₁	2·0	72 s	1·3	
	*114	i. t. (K)			50 d	5; 0·19; 0·55; 0·72	+4·7
	115	95·7 β⁻	114·904₁	0·6	6×10¹⁴ a	9/2	+5·51
	116	β⁻	115·905₆	3·3	14 s		
	*116	β⁻		<u>1·0</u>; 0·9; 0·6	54 min	5; 0·1 ... 2·1	+4·4
50 Sn	112	0·96	111·905₀			0	0
	113	K	112·905₀		119 d	0·26; ¹¹³In*	
	114	0·66	113·903₀			0	0
	115	0·35	114·903₅			½	−0·913
	116	14·30	115·902₁			0	0
	117	7·61	116·903₁			½	−0·995
	*117	i. t.			14 d	0·16	
	118	24·03	117·901₈			0	0
	119	8·58	118·903₄			½	−1·041
	*119	i. t.			245 d	3/2; 0·02; 0·07	+0·8
	120	32·85	119·902₁			0	0
	121	β⁻	120·904₂	0·38	27 h		
	*121	β⁻		0·42	>5 a		
	122	4·72	121·903₄			0	0
	123	β⁻	122·905₇	<u>1·4</u>; 0·4	125 d	1·1	
	*123	β⁻		1·3	40 min	0·15	
	124	5·94	123·905₂			0	0
	125	β⁻	124·907₈	<u>2·4</u>; 0·4	9·4 d	0·2 ... 1·9	
	*125	β⁻		<u>2·0</u>; 0·5	10 min	0·3 ... 1·9	
51 Sb	121	57·25	120·903₈			5/2	+3·34
	122	β⁻(K, β⁺)	121·905₁	2·0; 1·4; 0·7	2·8 d	2; 0·56...1·3	−1·9
	*122	i. t.			3·5 min	0·07; 0·06	

Element Z	A	$a[\%]$ or disint.	M u	E MeV	T	I or E_γ	μ
	123	42·75	$122 \cdot 904_2$			$\frac{7}{2}$;	+2·53
	124	β^-	$123 \cdot 905_9$	2·3..0·6..0·3	60 d	3; 0·6 ... 2·1	
	*124	i. t., β^-		3·2	1·3 min	0·01	
	*124	i. t., β^-		2·5	21 min	0·02	
	125	β^-	$124 \cdot 905_2$	0·6...0·1	2 a	0·04 ... 0·6	
52 Te	**120**	0·09	$119 \cdot 905$			0	0
	121	K			17 d	0·57	
	122	2·48	$121 \cdot 903_0$			0	0
	123	0·87 (K?)	$122 \cdot 904_2$		$> 5 \times 10^{13}$ a	$\frac{1}{2}$	−0·732
	*123	i. t.			104 d	0·09; 0·16	
	124	4·16	$123 \cdot 902_8$			0	0
	125	6·99	$124 \cdot 904_4$			$\frac{1}{2}$	−0·882
	*125	i. t.			58 d	0·04; 0·11	
	126	18·7	$125 \cdot 903\ 2$			0	0
	127	β^-	$126 \cdot 905\ 1$	0·7	9·3 h	0·06 ... 0·4	
	*127	i. t. (β^-)			105 d	0·09 (0·66)	
	128	31·8	$127 \cdot 904_7$			0	0
	129	β^-	$128 \cdot 906\ 5_8$	1·5	73 min	0·03 ... 1·1	
	*129	i. t.			33 d	0·11	
	130	34·5	$129 \cdot 906_7$			0	0
	131	β^-	$130 \cdot 908\ 5_8$	2·1; 1·7; 1·4	25 min	0·1 ... 1·1	
	*131	β^-, i. t.		2·5 ... 0·4	1·2 d	0·1 ... 1·1; 0·18	
53 I	125	K	$124 \cdot 904_6$		60 d	$\frac{5}{2}$; 0·035	3
	127	100	$126 \cdot 904\ 3_5$			$\frac{5}{2}$	2·79
	128	β^-, K	$127 \cdot 905\ 8_2$	2·1; 1·1	25·0 min	1; 0·4 ... 1·0	
	129	β^-	$128 \cdot 904\ 9_9$	0·15	$1 \cdot 6 \times 10^7$ a	$\frac{7}{2}$; 0·04	2·60
	131	β^-	$130 \cdot 906\ 1_3$	0·61; 0·33; 0·25	8·1 d	$\frac{7}{2}$; 0·08...0·7; 0·36	2·74
	135	β^-		1·4; 1·0; 0·5	6·7 h	$\frac{7}{2}$; 0·4 ... 1·8	
54 Xe	124	0·096	$123 \cdot 906_1$			0	0
	126	0·09	$125 \cdot 904\ 2$			0	0
	128	1·92	$127 \cdot 903\ 5_4$			0	0
	129	26·44	$128 \cdot 904\ 7_8$			$\frac{1}{2}$	−0·773
	130	4·08	$129 \cdot 903\ 5_1$			0	0
	131	21·18	$130 \cdot 905\ 0_9$			$\frac{3}{2}$	+0·687
	132	26·89	$131 \cdot 904\ 1_6$			0	0
	133	β^-	$132 \cdot 905_6$	0·35	5·3 d	0·081	
	*133	i.t.			2·3 d	0·23	
	134	10·44	$133 \cdot 905\ 4_0$			0	0
	135	β^-	$134 \cdot 909_0$	0·9; 0·55	9·1 h	0·25; 0·6	
	*135	i.t.			15 min	0·53	
	136	8·87	$135 \cdot 907\ 2_2$			0	0
	137	β^-		3·5	3·9 min		
55 Cs	131	K	$130 \cdot 905\ 4_7$		10 d	$\frac{5}{2}$	+3·5
	133	100	$132 \cdot 905_1$			$\frac{7}{2}$	+2·564
	134	β^-	$133 \cdot 906_5$	1·4...0·6... 0·09	2·2 a	4; 0·2 ... 1·4	+2·97
	*134	i.t. (β^-)			3·1 h	8; 0·01; 0·13; 0·14	+1·1

Element Z	A	$a[\%]$ or disint.	M u	E MeV	T	I or E_γ	μ
	135	β^-	$134 \cdot 905_8$	$0 \cdot 21$	2×10^6 a	$\frac{7}{2}$	$+2 \cdot 713$
	137	$\beta^- \to {}^{137}\text{Ba}^*$	$136 \cdot 906_8$	$1 \cdot 2;\ \underline{0 \cdot 51}$	28 a	$\frac{7}{2}$	$+2 \cdot 822$
56 Ba	130	$0 \cdot 101$	$129 \cdot 906\ 2_5$			0	0
	131	K			12 d	$0 \cdot 06 \ldots 1 \cdot 7$	
	132	$0 \cdot 097$	$131 \cdot 905$			0	0
	133	K	$132 \cdot 905_6$		7·5 a	$0 \cdot 05 \ldots 0 \cdot 38$	
	*133	i.t.			39h	$0 \cdot 01 \ldots 0 \cdot 28$	
	134	$2 \cdot 42$	$133 \cdot 904_3$			0	0
	135	$6 \cdot 59$	$134 \cdot 905_6$			$\frac{3}{2}$	$+0 \cdot 832$
	136	$7 \cdot 81$	$135 \cdot 904_4$			0	0
	137	$11 \cdot 32$	$136 \cdot 905_6$			$\frac{3}{2}$	$+0 \cdot 931$
	*137	i.t.			2·6 min	$0 \cdot 662$	
	138	$71 \cdot 66$	$137 \cdot 905_0$			0	0
	139	β^-	$138 \cdot 908_6$	$2 \cdot 4;\ \underline{2 \cdot 2};\ 0 \cdot 8$	85 min	$0 \cdot 16;\ 1 \cdot 4$	
	140	$\beta^- \to {}^{140}\text{La}$	$139 \cdot 910_5$	$\underline{1 \cdot 0} \ldots 0 \cdot 5$	12·8 d	$0 \cdot 03 \ldots 0 \cdot 54$	
57 La	138	$0 \cdot 089\ K,\ \beta^-$	$137 \cdot 906_8$	$0 \cdot 21$	$1 \cdot 1 \times 10^{11}$ a	$5;\ 1 \cdot 4;\ 0 \cdot 81$	$+3 \cdot 68$
	139	$99 \cdot 91$	$138 \cdot 906_1$			$\frac{7}{2}$	$+2 \cdot 761$
	140	β^-	$139 \cdot 909_3$	$2 \cdot 2 \ldots 0 \cdot 4$	40·2 h	$3;\ 0 \cdot 07 \ldots 2 \cdot 9$	
58 Ce	136	$0 \cdot 19$	$135 \cdot 907$			0	0
	138	$0 \cdot 25$	$137 \cdot 905_7$			0	0
	139	K	$138 \cdot 906_4$		140 d	$\frac{3}{2};\ 0 \cdot 166$	$0 \cdot 8$
	140	$88 \cdot 5$	$139 \cdot 905_3$			0	0
	141	β^-	$140 \cdot 908_0$	$0 \cdot 58;\ \underline{0 \cdot 44}$	33 d	$\frac{7}{2};\ 0 \cdot 15$	$0 \cdot 9$
	142	$11 \cdot 1\ \alpha$	$141 \cdot 909_0$	$1 \cdot 5$	5×10^{15} a	0	0
	143	β^-	$142 \cdot 921\ 2$	$1 \cdot 4 \ldots 0 \cdot 2$	33 h	$0 \cdot 06 \ldots 1 \cdot 1$	
	144	$\beta^- \to {}^{144}\text{Pr}$	$143 \cdot 913\ 4$	$\underline{0 \cdot 32};\ 0 \cdot 24$ $0 \cdot 18$	284 d	$0 \cdot 03 \ldots 0 \cdot 13$	
59 Pr	141	100	$140 \cdot 907\ 4$			$\frac{5}{2}$	$+3 \cdot 9$
	142	β^-	$141 \cdot 909\ 8$	$\underline{2 \cdot 15};\ 0 \cdot 6$	19 h	$2;\ 1 \cdot 57$	
	143	β^-	$142 \cdot 910\ 6$	$0 \cdot 93$	13·6 h		
	144	β^-	$143 \cdot 913\ 1$	$\underline{3 \cdot 0};\ 2 \cdot 3;\ 0 \cdot 8$	17·3 min	$0 \cdot 7;\ 1 \cdot 5;\ 2 \cdot 2$	
60 Nd	142	$27 \cdot 1$	$141 \cdot 907\ 5$			0	0
	143	$12 \cdot 2$	$142 \cdot 909\ 6$			$\frac{7}{2}$	$-1 \cdot 1$
	144	$23 \cdot 8\ \alpha$	$143 \cdot 909\ 9$	$1 \cdot 8$	5×10^{15} a	0	0
	145	$8 \cdot 30$	$144 \cdot 912_2$			$\frac{7}{2}$	$-0 \cdot 7$
	146	$17 \cdot 2$	$145 \cdot 912_7$			0	0
	147	β^-	$146 \cdot 915\ 8$	$0 \cdot 81 \ldots 0 \cdot 2$	11·9 d	$\frac{5}{2}$	$0 \cdot 6$
	148	$5 \cdot 73$	$147 \cdot 916_5$			0	0
	149	β^-	$148 \cdot 919_8$	$1 \cdot 5;\ \underline{1 \cdot 1};\ 0 \cdot 95$	2 h	$0 \cdot 65 \ldots 0 \cdot 03$	
	150	$5 \cdot 62$	$149 \cdot 920_7$			0	0
61 Pm	145	K	$144 \cdot 912_3$		18 a	$0 \cdot 067;\ 0 \cdot 073$	
	147	β^-	$146 \cdot 914\ 9$	$0 \cdot 22$	2·6 a	$\frac{7}{2};\ 0 \cdot 12$	$+2 \cdot 7$
62 Sm	144	$3 \cdot 1$	$143 \cdot 911_7$			0	0
	147	$15 \cdot 0\ \alpha$	$146 \cdot 914\ 6$	$2 \cdot 1$	$1 \cdot 3 \times 10^{11}$ a	$\frac{7}{2}$	$-0 \cdot 08$
	148	$11 \cdot 2$	$147 \cdot 914_6$			0	0
	149	$13 \cdot 8$	$148 \cdot 916_9$			$\frac{7}{2}$	$-0 \cdot 6$
	150	$7 \cdot 4$	$149 \cdot 917_0$			0	0
	151	β^-	$150 \cdot 919_7$	$0 \cdot 08$	~ 93 a	$0 \cdot 02$	
	152	$26 \cdot 7$	$151 \cdot 919_5$			0	0
	153	β^-	$152 \cdot 921_7$	$0 \cdot 80; 0 \cdot 7; 0 \cdot 6$	47 h	$\frac{3}{2};\ 0 \cdot 07 \ldots 0 \cdot 6$	

Element Z	A	$a[\%]$ or disint.	M u	E MeV	T	I or E_γ	μ
	154	22·7	153.922_0			*0*	0
	155	β^-	154.924_7	1·6	24 min	0·10;0·14;0·25	
63 Eu	**151**	47·8	150.919_6			$\frac{5}{2}$	3·6
	152	K, β^-, β^+	151.921_5	β^-: 1·5 ... 0·2	12·5 a	3; 0·1 ... 1·4	2·0
	*****152**	β^-, K, β^+		β^+: <u>1·9</u>; 1·6	9·3 h	0·1 ... 1·4	
	153	52·2	152.920_9			$\frac{5}{2}$	1·6
	154	β^-	153.922_8	1·9 ... 0·15	16 a	3, 0·1 ... 1·6	2·1
64 Gd	**152**	0·20a	151.919_5		10^{15} a	*0*	0
	153	K	152.921_1		200 d	0·08 ... 0·1	
	154	2·15	153.920_7			*0*	0
	155	14·7	154.922_6			$\frac{3}{2}$	−0·3
	156	20·5	155.922_1			*0*	0
	157	15·7	156.923_9			$\frac{3}{2}$	−0·4
	158	24·9	157.924_1			*0*	0
	159	β^-	158.926_0	<u>0·94</u>; 0·88; 0·6	18 h	$\frac{3}{2}$; 0·06 ... 0·36	
	160	21·9	159.927_1			*0*	0
	161	β^-	160.929_3	<u>1·6</u>; 1·5	3·7 min	0·06 ... 0·53	
65 Tb	**159**	100	158.925_0			$\frac{3}{2}$	~1·5
	160	β^-	159.926_8	1·7 ... 0·3	73 d	3; 0·06 ... 1·5	
66 Dy	**156**	0·05	155.923_8			*0*	0
	158	0·09	157.924_0			*0*	0
	160	2·29	159.924_8			*0*	0
	161	18·9	160.926_6			$\frac{5}{2}$	0·4
	162	25·5	161.926_5			*0*	0
	163	25·0	162.928_4			$\frac{5}{2}$	0·5
	164	28·2	163.928_8			*0*	0
	165	β^-	164.931_7	<u>1·3</u>; 1·2; 0·3 0·9	140 min	$\frac{7}{2}$; 0·04 ... 1·1	
	*****165**	i.t. (β^-)			1·2 min	0·11	
67 Ho	**165**	100	164.930_3			$\frac{7}{2}$	3·3
	166	β^-	165.932_4	1·84 ... 0·23	27 h	0·08 ... 1·6	
68 Er	**162**	0·14	161.938_8			*0*	
	164	1·56	163.929_3			*0*	0
	166	33·4	165.930_4			*0*	0
	167	22·9	166.932_1			$\frac{7}{2}$	0·5
	*****167**	i.t.			2·5 s	0·21	
	168	27·1	167.932_4			*0*	0
	169	β^-	168.934_7	<u>0·34</u>; 0·33	9 d	$\frac{1}{2}$; 0·008	
	170	14·9	169.935_5			*0*	0
	171	β^-	170.938_2	1·5; <u>1·1</u>	7·5 h	$\frac{5}{2}$; 0·005 ... 0·9	
69 Tm	**169**	100	168.934			$\frac{1}{2}$	−0·2
	170	β^- (K)	169.935_9	<u>0·97</u>; 0·88	129 d	*1*; 0·08	0·3
70 Yb	**168**	0·14	167.933_8			*0*	0
	169	K			32d	0·008 ... 0·31	
	170	3·1	169.934_9			*0*	0
	171	14·3	170.936_5			$\frac{1}{2}$	+0·5
	172	21·8	171.936_6			*0*	0
	173	16·1	172.939_0			$\frac{5}{2}$	−0·7
	174	31·8	173.939_0			*0*	0

Element Z	A	a[%] or disint.	M u	E MeV	T	I or E_γ	μ
	175	β⁻	174·941₄	0·47; 0·35; 0·07	4·2 d	0·11 ... 0·4	
	176	12·7	175·942₇			0	0
	177	β⁻	176·945₅	1·4 ... 0·2	2 h	0·12 ... 1·2	
71 Lu	175	97·4	174·940₉			7/2	+2·0
	176	2·6 β⁻	175·942₇	0·4	2×10¹⁰ a	7;0·09;0·2;0·3	+2·8
	*176	β⁻		1·2; 1·1	4 h	1; 0·09	
	177	β⁻	176·944₀	0·50 ... 0·18	6·8 d	7/2;0·07 ... 0·32	
72 Hf	174	0·18	173·940₃			0	0
	175	K			70 d	0·09 ... 0·43	
	176	5·2	175·941₄			0	0
	177	18·5	176·943₅			7/2	+0·6
	178	27·1	177·943₉			0	0
	179	13·8	178·946₀			9/2	−0·5
	180	35·2	179·946₈			0	0
	181	β⁻	180·949₁	1·0; 0·41	45 d	0·004 ... 0·48 ... 0·70	
73 Ta	180	0·012	179·947 5				
	181	99·99	180·948 0			7/2	2·34
	182	β⁻	181·950 1	0·51	115 d	0·03 ... 1·5	
	*182	i.t.			16 min	0·15 ... 0·36	
74 W	180	0·14 α	179·947 0		3 × 10¹⁴ a	0	0
	181	K	180·948 2	3	145 d	0·14	
	182	26·4	181·942 3			0	0
	183	14·4	182·950 3			1/2	+0·12
	184	30·6	183·951 0			0	0
	185	β⁻	184·953₅	0·43	74 d	0·125	
	*185	i.t.			1·7 min	0·07 ... 0·17	
	186	28·4	185·954 3			0	0
	187	β⁻	186·957 4	1·3; 0·6; 0·3	24 h	0·07 ... 0·87	
75 Re	185	37·1	184·953₀			5/2	+3·14
	186	β⁻, K	185·955₁	1·07; 0·93	90 h	0·14 ... 0·77	
	187	62·9 β⁻	186·956 0	0·04	10¹¹ a	5/2	+3·18
	188	β⁻	187·958₂	2·1; 2·0	17 h	0·16 ... 2·0	
76 Os	184	0·02	183·952₆			0	0
	186	1·6	185·954₀			0	0
	187	1·6	186·956 0			1/2	+0·07
	188	13·3	187·956₀			0	0
	189	16·1	188·958₃			3/2	+0·651
	190	26·4	189·958₆			0	0
	191	β⁻	190·961₂	0·14	15 d	0·07	
	*191	i.t.			14 h	0·07	
	192	41·0	191·961 4			0	0
	193	β⁻	192·964 5	1·14 ... 0·7	32 h	0·07 ... 0·56	
77 Ir	191	37·3	190·960 9			3/2	+0·16
	192		191·963 0	0·67; 0·54; 0·24	75 d	0·2 ... 1·4	
	*192	i.t.			1·4 min	0·06	
	193	62·7	192·963 3			3/2	+0·17
	194	β⁻	193·965 2	2·2 ... 0·5	20 h	0·3 ... 2·1	

Table of Isotopes (Cont.)

Element Z	A	a[%] or disint.	M u	E MeV	T	I or E_γ	μ
78 Pt	190	0.013 α	189.960_0	3.3	10^{12} a	0	0
	192	0.78 α	191.961 4	~2.6	~10^{15} a	0	0
	193	K	192.963_3		<500 a		
	*193	i.t.			4.4 d	0.013; 0.13	
	194	32.9	193.962 8			0	0
	195	33.8	$194.964\ 8_2$			$\frac{1}{2}$	+0.600
	196	25.3	$195.964\ 9_8$			0	0
	197	β^-	$196.967\ 3_6$	0.67; 0.48; 0.47	19 h	0.08; 0.19; 0.28; 0.34	
	*197	i.t.			83 min		
	198	7.2	197.967_5			0	0
	199	β^-	198.970_7	1.7 ... 0.8	30 min	0.07 ... 0.96	
79 Au	196	$K(\beta^-)$	$195.966\ 5_5$	0.3	5.6 d	0.33 (0.43)	+0.14
	197	100	$196.966\ 5_5$			$\frac{3}{2}$	
	198	β^-	$197.968\ 2_4$	(1.37); 0.96	2.7 d	2; 0.41	0.5
	199	β^-	$198.968\ 7_5$	(0.46); 0.30 0.25	3.15 d	$\frac{3}{2}$; 0.05 ... 0.21	0.2
80 Hg	196	0.15	$195.965\ 8_2$			0	0
	197	K			66 h	$\frac{1}{2}$; 0.08; 0.19	0.6
	*197	i.t. (K)			24 h	$\frac{13}{2}$; 0.13; 0.16	−1.0
	198	10.0	$197.966\ 7_7$			0	0
	199	16.8	$198.968\ 2_6$			$\frac{1}{2}$	+0.53
	200	23.1	$199.968\ 3_4$			0	0
	201	13.2	$200.970\ 3_2$			$\frac{3}{2}$	0.59
	202	29.8	$201.970\ 6_2$			0	0
	203	β^-	$202.972\ 8_5$	0.21	47 d	0.28	
	204	6.9	$203.973\ 4_8$			0	0
	205	β^-	204.976_2	1.6; 1.4	5.1 min	0.2	
81 Tl			201.972_1		12.5 d	2; 0.44	≤0.15
	203	29.5	202.972_3			$\frac{1}{2}$	+1.596
	204	β^-, K	$203.973\ 8_9$	0.76	~4 a	2; no γ	0.09
	205	70.5	$204.974\ 4_6$			$\frac{1}{2}$	+1.612
	205	β^-	$205.976\ 0_8$	1.6	4.2 min		
(AcC″)	207	β^-	$206.977\ 4_5$	1.44	4.8 min	0.87	
(ThC″)	208	β^-	$207.982\ 0_1$	(2.4); 1.8; 1.6; 1.2	3.1 min	0.04 ... 2.6	
(RaC″)	210	β^-	$209.990\ 0_0$	1.9	1.32 min	0.3 ... 2.4 .	
82 Pb	203	K	202.973 4		52 h	0.68	
	204	1.5α	$203.973\ 0_7$	2.6	1.4×10^{17}a	0	0
(RaG)	206	23.6	$205.974\ 4\ _6$			0	0
(AcD)	207	22.6	$206.975\ 9_0$			$\frac{1}{2}$	+0.584
(ThD)	208	52.3	$207.976\ 6_4$			0	0
	209	β^-	$208.981\ 0_9$	0.64	3.3 h		
(RaD)	210	β^-	$209.984\ 1_8$	0.06; 0.018	20 a	0.047	
(AcB)	211	β^-	210.988 8	1.39; 0.5	36.1 min	0.07 ... 0.8	
(ThB)	212	β^-	$211.991\ 9_0$	0.58; 0.34;	10.6 h	0.1 ... 0.4	
(RaB)	214	β^-	213.999 8	0.65; 0.59	26.8 min	0.05 ... 0.8	
83 Bi	209	100	$208.980\ 4_2$			$\frac{9}{2}$	+4.040
	210	$\alpha(\beta^-)$	$209.984\ 1_1$	α: 4.9	3×10^6 a		
(RaE)	*210	$\beta^-(\alpha)$		β^-; 1.17	5.0 d	1	~0

Element		a[%] or disint.	M u	E MeV	T	I or E$_\gamma$	μ
Z	A						
(AcC)	211	$\alpha(\beta^-)$	210·987 2$_9$	α; 6·6; 6·3	2·15 min	0·35	
(ThC)	212	β^-, α	211·991 2$_7$	β: 2·25 α:6·09; 6·05	60·5 min	0·1 ... 2·2	
(RaC)	214	$\beta^-(\alpha)$	213·998 2$_3$	β3·2; α5·5; 5·4	19·7 min	0·6 ... 2·4	
84 Po	209	α	208·982 4$_6$	4·88	200 a	$\frac{1}{2}$	
(RaF)	210	α	209·982 8$_7$	5·30	138 d	0·8	
(AcC')	211	α	210·986 6$_5$	7·44	0·6 s	0·9; 0·6	
(ThC')	212	α	211·988 8$_6$	8·78	3×10^{-7} s		
(RaC')	214	α	213·995 1$_9$	7·68	$1·6 \times 10^{-4}$ s		
(AcA)	215	$\alpha(\beta^-)$	214·999 5	α: 7·38	$1·8 \times 10^{-3}$ s		
(ThA)	216	$\alpha(\beta^-)$	216·001 9$_2$	α: 6·78	0·16 s		
(RaA)	218	$\alpha(\beta^-)$	218·008 9	α: 6·00	3·05 min		
85 At	210	$K(\alpha)$	209·987$_0$	5·52; 5·4	8·3 h	0·05 ... 2·6	
	215	α	214·998 6$_6$	8·0	$\sim 10^{-4}$ s		
	216	α	216·002 4$_0$	7·8	$\sim 3 \times 10^{-4}$ s		
	218	$\alpha(\beta^-)$	218·008 5$_5$	α: 6·7	~ 2 s		
86 Rn (Em)							
(An)	219	α	219·009 5$_2$	6·8; 6·5; 6·4	3·92 s	0·3; 0·4	
(Tn)	220	α	220·011 4$_0$	6·28	52 s	0·54	
(Rn)	222	α	222·017 5	5·48	3·825 d	0·51	
87 Fr							
(AcK)	223	$\beta^-(\alpha)$	223·019 8	β:1·2; α: 5·3	22 min	0·08; 0·22; 0·3	
88 Ra							
(AcX)	223	α	223·018 5$_6$	5·87 ... 5·71 ... 5·43	11·7 d	0·03 ... 0·45	
(ThX)	224	α	224·020 2$_2$	5·68; 5·45	3·64 d	0·24	
(Ra)	226	α	226·025 3$_6$	4·78; 4·60	1600 a	0; 0·19	0
(MsTh$_1$)	228	β^-	228·031 2$_3$	0·053	6·7 a	0	0
89 Ac	227	$\beta^-(\alpha)$	227·027 8$_1$	β0·046; α4·9	22 a	$\frac{3}{2}$	+1·1
(MsTh$_2$)	228	β^-	228·031 1$_7$	2·2 ... 0·5	6·13 h	0·06 ... 1·6	
90 Th							
(RdAc)	227	α	227·027 7$_7$	6·0 ... 5·7	18·2 d	0·03 ... 0·3	
(RdTh)	228	α	228·028 7$_5$	5·42; 5·34	1·91 a	0; 0·084; 0·13; 0·17; 0·22	0
	229	α	229·031 6$_3$	5·0; 4·9; 4·8	7×10^3 a	0·03 ... 0·3	
(Io)	230	α	230·033 0$_8$	4·68; 4·62	$8·0 \times 10^4$a	0; 0·07 ... 0·25	0
(UY)	231	β^-	231·036 3$_5$	0·3 ... 0·09	25·6 h	0·02 ... 0·3	
	232	100α	232·038 2$_1$	4·01; 3·95	$1·41 \times 10^{10}$ a		0
	233	β^-	233·041 4$_3$	1·23	22 min		
(UX$_1$)	234	β^-	234·043 5$_7$	0·19; 0·10	24·1 d	0·03 ... 0·09	
91 Pa	231	α	231·035 9$_4$	5·05 ... 4·67	$3·4 \times 10^4$ a	$\frac{3}{2}$; 0·03 ... 0·4	2·0
	233	β^-	233·040 1$_1$	0·57; 0·26; 0·15	27·4 d	$\frac{3}{2}$; 0·02 ... 0·42	+3·4
(UZ)	234	β^-	234·043 4	1·1; 0·5; 0·3	6·7 h	0·04 ... 1·7	
(UX$_2$)	*234	β^-(i.t.)		2·3; 1·5; 0·6	1·2 min	0·04 ... 1·8	
92 U	233	α	233·039 5$_0$	4·82	$1·6 \times 10^5$ a	$\frac{5}{2}$; 0·04 ... 0·1	+0·5
(U II)	234	0·005 6 α	234·040 9$_0$	4·77; 4·71	$2·5 \times 10^5$a	0; 0·05 ... 0·1	0
(AcU)	235	0·720 α	235·043 9$_3$	4·56 ... 4·38 ... 4·12	$7·1 \times 10^8$ a	$\frac{7}{2}$; 0·07 ... 0·4	−0·3

Element Z A	a[%] or disint.	M u	E MeV	T	I or E_γ	μ
236	α	$236.045\,7_3$	4.50	2.4×10^7 a	0.05	
237	β^-	$237.048\,5_8$	0.25	6.8 d	$0.03\ \ldots\ 0.37$	
(U I) 238	$99.27\ \alpha$	$238.050\,7_6$	$\underline{4.19};\ 4.14$	4.5×10^9 a	0.048	
239	β^-	$239.054\,3_2$	1.2	23.5 min	0.074	
93 Np 237	α	$237.048\,0_3$	$4.9\ \ldots\ 4.5$	2.2×10^6 a	$\tfrac{5}{2};\ 0.03\ldots0.20$	-8.5
238	β^-	$238.050\,9$	$1.2\ \ldots\ 0.3$	2 d	$2;\ 0.04\ \ldots\ 1.0$	
239	β^-	$239.052\,9_4$	$0.72\ \ldots\ 0.3$	2.3 d	$\tfrac{1}{2};\ 0.05\ \ldots\ 0.33$	
94 Pu 239	α	$239.052\,1_6$	$5.15\ \ldots\ 4.9$	2.4×10^4 a	$\tfrac{1}{2};\ 0.04\ \ldots0.42$	0.02
240	α	$240.053\,9_7$	$\underline{5.16};\ 5.12$	6.6×10^3 a	0.045	
241	$\beta^-(\alpha)$	$241.056\,7_1$	$\beta0.02;\ \alpha4.9$	13 a	$\tfrac{5}{2};\ 0\ (0.1)$	0.11
242	α	$242.058\,7$	4.90	3.8×10^5 a		
95 Am 241	α	$241.056\,6_9$	$5.5\ \ldots\ 5.3$	460 a	$\tfrac{5}{2};\ 0.03\ \ldots\ 0.37$	$+1.4$
242	$\beta^-,\ K$	$242.059\,4_8$	0.6	~100 a		
*242	$\beta^-,\ K$		$0.67;\ 0.63$	16 h	$1;\ 0.45;\ 0.042$	
243	α	$243.061\,3_8$	$5.34\ \ldots\ 5.17$	8×10^3 a	$\tfrac{5}{2};\ 0.075$	$+1.4$
96 Cm 242	α	$242.058\,8_0$	$\underline{6.11};\ 6.07$	163 d	$0.04\ \ldots\ 1.0$	
243	$\alpha(K)$	$243.061\,3_8$	$\underline{6.06}\ldots 5.63$	35 a	$0.21;0.23;0.28$	
244	α	$244.062\,9_1$	$\underline{5.80};\ 5.76$	18 a	$0.04;0.10;0.15$	
245	α	$245.065\,3_4$	$\underline{5.45};\ \underline{5.36}$	1×10^4 a	$0.13;\ 0.17$	
248	α, fis. Fis.		5.0	5×10^5 a		
97 Bk 243	$K(\alpha)$	$243.062\,9_2$	$6.72;\ \underline{6.55};$ 6.20	4.5 h	$0.04\ \ldots\ 0.54$	
245	$K(\alpha)$	$245.066\,2_4$	$6.37;\ 6.17;$ 5.89	5.0 d	$0.16\ \ldots\ 0.48$	
247	α	$247.070\,1_8$	$5.67;\ \underline{5.51};$ 5.30	10^4 a	$0.08;\ 0.27$	
249	$\beta^-(\alpha)$	$249.074\,8_4$	$\beta0.1;$ $\alpha5.4;\ 5.0$	310 d	0.32	
250	β^-	250.078_5	$1.9;\ 0.9$	3.2 h	1	
98 Cf 246	α	$246.068\,7_8$	$\underline{6.75};\ 6.71$	36 h	$0.04;\ 0.10;$ 0.15	
248	α	$248.072\,3_5$	6.3	~300 d		
249	α	$249.075\,7_0$	$6.2;\ 5.9;\ \underline{5.8}$	360 a	$0.05\ \ldots\ 0.34;$ 0.40	
250	α	$250.076\,5_5$	$\underline{6.02};\ 5.98$	10 a	0.043	
252	α (fis. Fis.)		$\underline{6.11};\ 6.07$	2.6 a	$0.043;\ 0.10$	
254	fis. Fis.			~60 d		
99 Es 251	$K(\alpha)$	$251.079\,8_5$	(6.5)	1.5 d		
253	α	$253.084\,6_8$	$\underline{6.63};\ 6.59$ $\ldots\ 6.18$	20 d	$0.04\ \ldots\ 0.43$	
254	$\beta^-(K,\alpha)$	254.088_1	$\beta1.0;\ \alpha6.4$	38 h	0.66	
100 Fm 250	$\alpha,\ K$	$250.079\,4_8$	7.4	30 min		
252	α	$252.082\,6_5$	7.0	30 h		
253	$K,\ \alpha$		6.9	~5 d		
254	α	$254.087\,0_0$	7.2	3 h	$0.04;\ 0.10$	
255	α		7.0	21 h	$0.06;\ 0.08$	
256	fis. Fis.			3 h		
101 Md 255	$K,\ \alpha$	255.090_6	7.3	0.5 h		
102 No 253	α		8.5	~10 min		
254	α		8.8	3 s		

This is a complex and rapidly changing subject. Since the discovery of the first mesons in 1937, a great number of other particles have been found, and the whole field of particle physics and resonant states is still under constant review. The table which follows contains data on the so-called 'stable' particles, *i.e.* those particles which are immune to decay via the strong interaction. The rest energy of each particle is given in units of MeV, to convert to other units, use may be made of the table of energy equivalents (p.p. 80-1).

FUNDAMENTAL PARTICLES

	Name	Symbol	Rest Energy M_0/MeV	Mean lifetime τ/s	Common decay modes
Leptons	Photon	γ	0	stable	
	Neutrino	ν_e	0	stable	
		ν_μ	0	stable	
	Electron	$e^\pm$	0·511 004(2)	stable	
	Muon	$\mu^\pm$	105·659(2)	$2\cdot1994(6) \times 10^{-6}$	$e\nu\bar{\nu}$
Mesons	Pion	$\pi^\pm$	139·576(11)	$2\cdot602(2) \times 10^{-8}$	$\mu\nu$
		π^0	134·972(12)	$0\cdot84(10) \times 10^{-16}$	$\gamma\gamma(99\%)\gamma e^+e^-(1\%)$
	Kaon	$K^\pm$	493·82(11)	$1\cdot235(4) \times 10^{-8}$	$\mu\nu(64\%)\pi^\pm\pi^0(21\%)$ $3\pi(5\%)$
		K^0	497·76(16)	$50\% K_1, 50\% K_2$	
		K_1		$8\cdot62(6) \times 10^{-11}$	$\pi^+\pi^-(69\%)2\pi^0(31\%)$
		K_2		$5\cdot38(19) \times 10^{-8}$	$\pi e\nu(39\%)\pi\mu\nu(27\%)$ $3\pi^0(21\%)\pi^+\pi^-\pi^0$ (13%)
	Eta	η^0	548·8(6)		$\gamma\gamma(38\%)\pi\gamma\gamma(2\%)3\pi^0$ $(31\%)\pi^+\pi^-\pi^0(23\%)$ $\pi^+\pi^-\gamma(5\%)$
Baryons	Proton	$p^\pm$	938·256(5)	stable	
	Neutron	n	939·550(5)	$9\cdot32(14) \times 10^2$	$pe\nu$
	Lambda	Λ^0	1115·60(8)	$2\cdot51(3) \times 10^{-10}$	$p\pi^-(65\%)n\pi^0(35\%)$
	Sigma	Σ^+	1189·4(2)	$8\cdot02(7) \times 10^{-11}$	$p\pi^0(52\%)n\pi^+(48\%)$
		Σ^0	1192·46(12)	$< 10^{-14}$	$\Lambda\gamma$
		Σ^-	1197·32(11)	$1\cdot49 \times 10^{-10}$	$n\pi^-$
	Xi	Ξ^0	1314·7(7)	$3\cdot03(18) \times 10^{-10}$	$\Lambda\pi^0$
		Ξ^-	1321·25(18)	$1\cdot66(4) \times 10^{-10}$	$\Lambda\pi^-$
	Omega	Ω^-	1672·5(5)	$1\cdot3(4) \times 10^{-10}$	$\Xi^0\pi^-, \Xi^-\pi^0, \Lambda K^-(?)$

29 Colour Codes in Electronics

Resistors

The colours on resistors are used to indicate the nominal value of their resistances, and the permitted tolerance on that value. In the *colour band system*, the resistor has three or four bands on it. The band at the end of the resistor indicates the first digit, the next band (working towards the centre of the resistor) indicates the second digit while the third band indicates the number of zeros which follow the two previous digits. The fourth band is used to indicate the manufacturers tolerance.

Some resistors are marked by the *body, tip and dot system* in which the first digit is indicated by the colour of the body of the resistor, the second digit by the band at one end of the resistor, and the number of zeros, by the band, or dot, in the centre of the resistor.

The colours used are as follows:

0	Black
1	Brown
2	Red
3	Orange
4	Yellow
5	Green
6	Blue
7	Violet
8	Grey
9	White

Preferred values of resistors (First two significant figs.)		
20% (no band)	10% (silver band)	5% (gold band)
10	10	10
		11
	12	12
		13
15	15	15
		16
	18	18
		20
22	22	22
		24
	27	27
		30
33	33	33
		36
	39	39
		43
47	47	47
		51
	56	56
		62
68	68	68
		75
	82	82
		91
100	100	100

Fuses

These are often marked by coloured dots on the glass of the fuse. The rating of the fuse is given by the following code:

60 mA	Black	1·0 A	Dark blue
100 mA	Grey	1·5 A	Light blue
150 mA	Red	2·0 A	Purple
250 mA	Brown	3·0 A	White
500 mA	Yellow	5·0 A	Black and white
750 mA	Green		

30 The Fundamental Constants

Certain physical constants have special importance on account of their universality or place in fundamental theory. These are given below, first in SI and then in cgs units.

The figure in brackets which follows the final digit, is the estimated uncertainty in the last digit.
Thus $c = 2{\cdot}997\ 924\ 590(8) \times 10^8$ m s^{-1} could be written $c = (2{\cdot}997\ 924\ 590 \pm 0{\cdot}000\ 000\ 008) \times 10^8$ m s^{-1}.

	Symbol	Quantity	Value	Multiplier and units — SI	Multiplier and units — cgs
General constants	c	Speed of light in vacuo	2·997 924 590(8)	10^8 m s^{-1}	10^{10} cm s^{-1}
	μ_0	Permeability of free space	4π	10^{-7} H m^{-1}	—
	ε_0	Permittivity of free space	8·854 19(1)	10^{-12} F m^{-1}	—
	e	Elementary charge	1·602 192(7)	10^{-19} C	10^{-20} e.m.u.
			or 4·803 25(2)	—	10^{-10} e.s.u.
	h	Planck's constant	6·626 20(5)	10^{-34} J s	10^{-27} erg s
	$h/2\pi$		1·054 592(8)	10^{-34} J s	10^{-27} erg s
	h/e	Quantum charge ratio	4·135 708(14)	10^{-15} J s C^{-1}	10^{-7} e.m.u.
			or 1·379 523(5)	—	10^{-17} e.s.u.
	α	Fine structure constant $= \dfrac{e^2}{2h\varepsilon_0 c}$	7·297 351(11)	10^{-3}	10^{-3}
	$1/\alpha$		1·370 360(2)	10^2	10^2
	G	Gravitational constant	6·673(3)	10^{-11} N m^2 kg^{-2}	10^{-8} dyn cm^2 g^{-2}
	Z_0	Impedance of free space	3·767 304(1)	10^2 Ohm	10^{11} e.m.u.
Electron	m_e	Electron rest mass	9·109 56(5)	10^{-31} kg	10^{-28} g
	$m_e c^2$	Electron rest energy	8·187 26(6)	10^{-14} J	10^{-7} erg
			or 5·110 041(16)	10^{-1} MeV	
	e/m_e	Electron charge-mass ratio	1·758 803(5)	10^{11} C kg^{-1}	10^7 e.m.u.
			or 5·272 759(16)	—	10^{17} e.s.u.
	λ_c	Compton wave length of electron	2·426 310(7)	10^{-12} m	10^{-10} cm
	r_e	Classical radius of electron	2·817 939(13)	10^{-15} m	10^{-13} cm